Lecture Notes
in Business Information Processing   258

More information about this series at http://www.springer.com/series/7911

Vijayan Sugumaran · Victoria Yoon
Michael J. Shaw (Eds.)

E-Life: Web-Enabled Convergence of Commerce, Work, and Social Life

14th Workshop on e-Business, WEB 2015
Fort Worth, Texas, USA, December 12, 2015
Revised Selected Papers

 Springer

Editors
Vijayan Sugumaran
Department of Decision and Information
  Sciences
Oakland University
Rochester, MI
USA

Michael J. Shaw
Beckman Institute for Advanced Science and
  Technology
University of Illinois at Urbana–Champaign
Urbana, IL
USA

Victoria Yoon
Virginia Commonwealth University
Richmond, VA
USA

ISSN 1865-1348   ISSN 1865-1356   (electronic)
Lecture Notes in Business Information Processing
ISBN 978-3-319-45407-8   ISBN 978-3-319-45408-5   (eBook)
DOI 10.1007/978-3-319-45408-5

Library of Congress Control Number: 2016949604

Printed on acid-free paper

This Springer imprint is published by Springer Nature
The registered company is Springer International Publishing AG Switzerland

Preface

The Workshop on e-Business (WeB) is a premier annual workshop on e-business and e-commerce. The purpose of the workshop is to provide an open forum for e-business researchers and practitioners worldwide to explore and respond to the challenges of next-generation e-business systems, share the latest research findings, explore novel ideas, discuss success stories and lessons learned, map out major challenges, and collectively chart the future directions of e-business. Since its inception in 2000, the WeB workshop has attracted state-of-the-art research and followed closely the developments in the technical and managerial aspects of e-business. The 14th Annual Workshop on e-Business (WeB 2015) was held in Fort Worth, Texas, on December 12, 2015. The workshop provided an interactive forum by bringing together researchers and practitioners worldwide to explore the latest challenges of next-generation e-business systems and the potential of service computing and big data analytics. Original research articles addressing a broad coverage of technical, managerial, economic, and strategic issues related to e-business, with emphasis on service computing and big data analytics, were presented at the workshop. These articles employed various IS research methods such as case study, survey, analytical modeling, experiments, computational models, design science, etc.

The theme of WeB 2015 was "Leveraging Service Computing and Big Data Analytics for E-Commerce." With the advances in high-speed connectivity and ability to deliver services seamlessly over the Web, service computing is not only reshaping how businesses operate and create new business solutions, but also transforming how e-business systems are designed, developed, and deployed. The global nature of service computing presents various opportunities and challenges, as well as creating a new economic structure for supporting different e-business models. At the same time, in this connected world, there is an explosion in the amount of data being created and collected, and big data analytics is increasingly being used by organizations to gain competitive advantage. Big data is changing the face of e-commerce by impacting all aspects of the business, be it inventory management, pricing, customer relationship management, new product/service innovation, and meeting customer demands. Big data analytics initiatives require considerable investment on infrastructure, tools, and technologies. While large organizations with adequate resources are able to capitalize on big data analytics, smaller companies are at a disadvantage. However, by providing big data analytics as Web services, companies of any size can gain access to these services and utilize them to support real-time decision-making, manage fraud, optimize pricing, and provide better customer service. Thus, leveraging service computing and big data analytics has great potential in transforming e-business operations for large and small businesses alike. WeB 2015 provided a forum for scholars to exchange ideas and share results from their research on service computing and big data analytics for e-commerce.

We received 45 submissions and each submission was reviewed by three reviewers. The Program Committee co-chairs had a final consultation meeting to look at all the reviews and make the final decisions on the papers to be accepted. We accepted 12 papers (26.7 %) as long/regular papers and 16 short papers.

We would like to thank all the reviewers for their time, effort, and completing their review assignments on time despite tight deadlines. Many thanks to the authors for their contributions.

January 2016

Vijayan Sugumaran
Victoria Yoon
Michael J. Shaw

The original version of the book was revised:
In the original publication of this book the
conference numbering was incorrect.
In the subtitle "15th Workshop on e-Business"
has now been corrected to "14th Workshop
on e-Business". The erratum to the
book is available at
https://doi.org/10.1007/978-3-319-45408-5_29

Organization

Honorary Chair

Andrew B. Whinston   University of Texas at Austin, USA

Conference Chair

Michael J. Shaw   University of Illinois at Urbana-Champaign, USA

Program Co-chairs

Vijayan Sugumaran   Oakland University, USA
Victoria Yoon   Virginia Commonwealth University, USA

Local Arrangements Chairs

Dan Kim   University of North Texas, USA
Gene Moo Lee   University of Texas at Arlington, USA

Program Committee

Sinan Aral   Massachusetts Institute of Technology, USA
Subhajyoti Bandyopadhyay   University of Florida, USA
Joseph Barjis   Delft University of Technology, The Netherlands
Peter Bernus   Griffith University, Australia
David Bodoff   University of Haifa, Israel
Indranil Bose   Indian Institute of Management, Calcutta, India
Ulrich Bretschneider   Kassel University, Hessen, Germany
Hsinlu Chang   National Chengchi University, Taiwan
Michael Chau   The University of Hong Kong, SAR China
Patrick Y.K. Chau   The University of Hong Kong, SAR China
Hsinchun Chen   University of Arizona, USA
Kenny Cheng   University of Florida, USA
Ching-Chin Chern   National Taiwan University
Romilla Chowdhuri   University of Massachusetts at Boston, USA
Honghui Deng   University of Nevada, Las Vegas, USA
Kutsal Dogan   Ozyegin University, Istanbul, Turkey
Aidan Duane   Waterford Institute of Technology, Ireland
Xianjun Geng   University of Texas, Dallas, USA
Hong Guo   University of Notre Dame, USA
Susanna Ho   Australia National University

Contents

Sentiment Analysis of Twitter Users Over Time: The Case of the Boston Bombing Tragedy

Jaeung Lee[1](✉), Basma Abdul Rehman[2], Manish Agrawal[3], and H. Raghav Rao[4]

[1] Department of Computer Information Systems,
College of Business, Louisiana Tech University,
P.O. Box 10318, Ruston, LA 71272, USA
mgmtjake@gmail.com
[2] Department of Management Science and Systems,
School of Management, State University of New York at Buffalo,
325 Jacobs Management Center, Buffalo, NY 14260, USA
[3] Department of Information Systems Decision Sciences,
Muma College of Business, University of South Florida,
4202 E. Fowler Avenue, BSN 3403, Tampa, FL 33620, USA
[4] Department of Information Systems and Cyber Security,
College of Business, The University of Texas at San Antonio,
One UTSA Circle, San Antonio, TX 78249, USA

Abstract. Social Network Services (SNS), for example Twitter, play a significant role in the way people share their emotions about specific events. Emotions can spread via SNS and can spur people's future actions. Therefore, during extreme events, disaster response agencies need to manage emotions appropriately via SNS. In this research, we investigate the Twitter verse associated with an event - the Boston Bombing context. We focus on tweets in the context of hazard-describing keywords (Explosion, Bomb), important event timelines, and the related changes in emotions over time. We compare the results with a corpus of tweets collected at the same time that are not associated with the above hazard- describing keywords. A sentiment analysis shows anger was the most strongly expressed emotion in both groups. However, there were statistical differences in Anxiety and Sadness among the two groups over time.

Keywords: Social media · Big data · Emotion spread · Announcement · Disaster response

1 Introduction

The emergence of Social Network Services (SNS) has provided an additional option for people to share their personal emotions regarding specific events. Before the emergence of SNS, due to the restriction of communication channels, people could primarily share their emotions while they were physically close to people such as family members or friends via traditional media (phone, letter, or personal face to face meetings). These

© Springer International Publishing Switzerland 2016
V. Sugumaran et al. (Eds.): WEB 2015, LNBIP 258, pp. 1–14, 2016.
DOI: 10.1007/978-3-319-45408-5_1

traditional communication channels limited the diffusion of emotions. The advent of SNS has provided new opportunities that enable people to express their emotions without any time or location restrictions and have allowed the spread of expressed emotions over a much larger footprint.

In the SNS platform, people's emotions can be expressed via SNS postings. These postings that include people's emotions have been found to be significant factors in various social actions. For instance, prior literature in this field has studied the impact of emotions on the public in the context of various disaster events. According to Pacherie [1], emotions are often portrayed as motivational forces of action or as the stimuli for action. Emotions that diffuse through SNS can also spur the future action of users. This is one of the main reasons why managing emotions in SNS is important and needs to be handled appropriately.

Especially, when people express negative emotions such as anger, sadness and anxiety, the mitigation of such emotions is important for disaster response authorities. This is because propagated negative emotions can spur unexpected negative actions. The motivation of this study is to understand the various characteristics of SNS messages during extreme events, such as emotion differences between different types of related tweets, the impact of announcements on emotions, the impact of announcements or news on expressed emotions. We are also interested in understanding the pervasive emotions during the extreme event, in order to provide valuable information to disaster response authorities. By using the results of this study, disaster response authorities can develop plans to mitigate public chaos and prepare for unexpected future behaviors by citizens.

The literature on emotion contagion in online communication [2, 3], has documented that the emotions of online communication users are influenced by online posts or messages of their online friends. Kramer [4] has addressed this emotion contagion in the context of SNS - when an SNS user posts a message, words used by him/her influence later word selection by their friends. That is, people's emotions could also be affected by the words used by SNS users. The literature on social information processing points out that in text based interaction, people may describe their thoughts, emotions, or attitudes by selecting their word and punctuation use [5].

Because of the characteristics of SNS, diffusion of emotions in the SNS environment is faster and broader than that of traditional communication channels [6]. The implication is that during extreme events, if related disaster response authorities such as police departments, and law enforcement agencies cannot deal with the spread of emotion in a timely manner, consequences of this diffusion can be much more difficult to manage. Therefore, investigating expressed negative emotions is an important issue [7]. Back et al. [8] studied emotion changes through the timeline of the World Trade Center terrorist attack. They explained that announcements and news about the incident impacted people's expressed emotions which were distributed by text pagers. Around a decade later, people are using SNS instead of text pagers. SNS increase accessibility and diversity of information about any event.

In this research, we explore negative emotions (anger, anxiety, and sadness) expressed in Twitter space during one extreme event - the Boston bombing tragedy of 2013. Our research questions are (1) Are there emotional differences between types of tweet that are specifically related to the incident and those that are not specifically related to the incident? (2) Do incident related announcements or news have an impact on people's emotions? (3) What type of negative emotions are presented more during terrorist attacks?

In order to answer our research questions, we focused on the Twitter messages that were created by Twitter users and captured during the incident. Many news media channels broadcasted the actual situation about the Boston tragedy in real time and people could obtain and share real time information via SNS. We assume that during the 5 days of the event period (from the time that the bombing happened to the time that the suspect was captured), announcements or news regarding the incident impacted people's expressed emotion on SNS.

We separated the Boston bombing related Twitterverse into two spheres: one which focused on messages specifically related to the bombing and the explosion (Include the keywords "bomb" and "explosion"), while the second concentrated on messages not specifically focused on the incident but collected at the same time and which were related to Boston or the Boston marathon incident. In addition, by analyzing tweet messages collected hourly, we studied the impact of event-related announcements or news on expressed negative emotions (anger, anxiety, and sadness) as displayed on SNS.

The rest of this paper is organized as follows. First, we discuss the background and related literatures. Second, we explain the technique we used for sentiment analysis of emotions, Next, we introduce the data sets used in this research and describe the analysis and results. Finally, we conclude with a discussion of the potential implications and contributions of our research.

2 Research Background and Literature Review

2.1 Boston Marathon Bombing Tragedy

The Department of Homeland Security (DHS) reported that at 2:49 PM on April 15, 2013, two pressure cooker bombs (Improvised Explosive Device) placed near the 2013 Boston Marathon finish line exploded within seconds of each other, resulting in a death toll of three and injuries to more than two hundred people. On April 18th, the Federal Bureau of Investigation (FBI) identified Tamerlan Tsarnaev and Dzhokhar Tsarnaev as the primary suspects in this tragedy and released photographs and surveillance videos of them. One suspect, Tamerlan Tsarnaev, was shot during his encounter with the police and died shortly in Watertown, Massachusetts. The other suspect, Dzhokhar Tsarnaev, who ran away from the scene was caught the following day at 8:50 PM on April 19th [9].

When the explosions occurred near the finish line, around 5,700 runners were still in the race. Within 1 min, the event had been announced on SNS. During the period (April 15 – 19), many messages about the bombing incident were propagated via SNS, especially on Twitter. The messages created benefits as well as confusion among the public. There were a myriad of tweets that mentioned the incident itself, and gave actionable information that provided real time information to the public, information about tracking suspects providing useful clues. However, they also contributed to the apprehensions of law enforcement agencies because of the messages, which spread negative emotions such as anger, anxiety, and sadness to the public. Moreover, as these tweets were diffused the embedded emotions in the messages were also propagated.

2.2   Emotions

Many English language emotion analysis studies have applied Ekman's 6 basic emotion types [10–12] to classify people's sentiment from text based documents [13, 14]. The 6 basic emotions consist of Surprise, Happiness, Anger, Fear, Disgust, and Sadness. In this paper we focus on the three negative emotions, anger, fear and sadness: Anger is an emotion that includes an uncomfortable response to a perceived (or real) grievance [15].; Fear can be explained as being afraid of something and this is caused by being aware of danger [16]; sadness is the opposite emotion of happiness [17]. We suggest that the three negative emotions Anger, Fear (Anxiety), and Sadness help in understanding the tragedy better. We do not consider "disgust" among the list of emotions to study - "feelings of disgust are often immune to rationality" [18].

2.3   Emotional Contagion

"Emotional contagion" is described as "a process in which a person or group influences the emotion or behavior of another person or group through the conscious or unconscious induction of emotional states and behavioral attitudes [19]." Recently, Vijayalakshmi and Bhattacharyya [20] pointed out that emotions and emotional contagion are being increasingly considered as important factors which influence individual behavior. Burke et al. [21] mention that people's interactions within the SNS can mirror the manner in which people interact with others in their offline lives. Moreover, in the SNS, people need to pay more attention to emotions and emotional contagion than in the offline environment. As we mentioned earlier, development of SNS eliminates time and location restrictions and allows people to express their emotions across a much wider audience. The degree of negative emotion contagion via SNS is also uncontrollable if disaster response authorities cannot manage contagious emotions in a swift manner. Therefore, understanding the pattern of emotion changes and the impact of external factors such as announcements or news during the extreme event situation is essential.

2.4   Warning Sign Words

According to Ma et al. [22] signal words have a relationship to a person's emotional levels (perceived hazard level). When there is a human participant in the context of specific signal word, the emotion of the participant regarding an action-related target is different. For example, when participants played in a shooting game, emotions of the participants that were told about armed targets compared to unarmed targets were different [23]. In addition, when people were showed words which described risky environmental events, their psychological response was different [24]. This indicates that the use of keywords (directly related to incident) in the SNS messages can reflect on peoples' emotions and may show different levels of negative emotions regarding the specific event compared to when messages do not contain event related keywords.

3 Methodology

3.1 Data Collection

On April 15th 2013, the race day turned deadly for Boston as two bombs went off near the finish line of the marathon, killing 2 spectators and injuring more than 260 people. Almost 4 h after the occurrence of this incident, we started collecting the data from Twitter using a custom tool developed for Twitter data collection. This tool uses the Twitter Streaming Application Program Interface (Twitter API) that helped us collect live feeds from Twitter. There were three keywords that were particularly of interest to us - "Boston", "BostonMarathon" and "BostonBombing". The search for Twitter feeds was based only on these three keywords. The data collection for the four-hour window that was missed out on initially (from 15th April 2013 18:52 GMT to 15th April 2013 21:53 GMT) was captured using a third party software called Topsy. Using Topsy, a hashtag-based search was performed using the same three keywords as mentioned above. This was to ensure uniformity in the data collection.

A total of 1,149,678 tweets were collected for this five-day period. The data was then separated into 2 major categories: one that included the keywords "explosion" and/or "bomb" (henceforth referred to as EB set), the other which did not include the keywords (henceforth referred to as non-EB set). For the simplification of the analysis of the data, we further divided the dataset in an hourly fashion (108 h which encompasses our 5 day dataset). Table 1 below indicates the details of the count of tweets by day:

Table 1. Message description

	4/15	4/16	4/17	4/18	4/19	Total number of tweets
(1) Number of tweets (with keywords EB)	146,601 (12.75%)	139,431 (12.13%)	48,278 (4.20%)	77,450 (6.74%)	43,239 (3.76%)	454,999 (39.58%)
(2) Number of tweets (without keyword - nonEB)	61,328 (5.33%)	391,169 (34.02%)	63,737 (5.54%)	109,154 (9.50%)	69,291 (6.03%)	694,679 (60.42%)
(1) + (2)	207,929 (18.09%)	530,600 (46.15%)	112,015 (9.74%)	186,604 (16.23%)	112,530 (9.79%)	1,149,678 (100%)

3 emotions were analyzed in the tweets – Sadness, Anger and Anxiety. We used Linguistic Inquiry and Word Count (LIWC) based sentiment analysis to chart out graphs for the three emotions mentioned and catalogued them separately for the EB and

non-EB datasets. The hourly transition of tweets for the 108 h period was the next step in data collection. For each emotion and every hour, we analyzed the transition of emotion and classified it either as peak (rise in value) or trough (decline in value). Peaks and troughs were based on the delta values (change in value of an emotion). This was carried out for the entire 108 h period, for the five day period and for all three emotions. Delta values of only 40 % and above were considered for the non-EB dataset and 50 % and above were considered for the EB dataset. Major events of every day during the 5-day period of the Boston Bombing incident were identified using these delta values. The content analysis was then performed taking into consideration the delta values and the time and event corresponding to these delta values.

3.2 Sentiment Analysis

In order to investigate the impact of keywords (use of Bomb or Explosion) directly related to the Boston bombing tragedy on people's expressed emotions, we conducted sentiment analysis using hourly collected tweet messages. The vital task carried out by sentiment analysis is to identify how sentiments are expressed in the texts or messages [25]. According to Stieglitz and Krüger [26], "this method is based on natural language processing, computational linguistics, and text analytics to identify and extract sub-jective information in different kinds of source materials."

In order to extract the sentiment of tweets automatically, we selected the LIWC 2007 software [27], because the LIWC based sentiment analysis has been previously conducted to analyze conversations that use instant messages, articles, or twitter messages [28–30]. LIWC is a text analysis software which has been developed to measure inherent emotions in the text using a psychometrically validated dictionary. This provides us with the rate of negative emotion word use in the messages by calculating their relative frequency based on the categorized dictionary.

For the three selected emotion types: anger, anxiety, and sadness, we captured the significant emotional changes in both EB and non-EB groups. In the process, we applied thresholds of more than 50 % of emotion change for EB and more than 40 % of emotional change for non-EB to analyze significant changes. Based on the results of sentiment analysis, we explored the expressed negative emotional differences between EB and non-EB groups in hourly time slices.

3.3 Content Analysis

Based on the published timelines of the Boston Bombing tragedy [31, 32], we per-formed content analysis. Two English speaking graduate students analyzed tweet data that was separated as EB and non-EB. We selected all announcements about the bombing itself as a major event on all days of EB and non-EB groups. For example, we decided on the following event as an announcement on day 2 for group of EB.

2 PM – 8 PM: FBI bulletin states that bombs were made by packing explosives, shrapnel and nails into pressure cookers. They are still unaware as to who detonated them and why.

The Number of Keywords used in the 2nd day's group of EB tweet messages.

FBI – 3353, Shrapnel – 36861, Pressure cooker – 6665, Detonate – 248, Bomb – 140471

Subsequently, using content analysis results (from EB and non-EB), we analyzed hourly sentiment analysis results to investigate the impact of the announcement on expressed emotion change in SNS.

4 Analysis Result

In this section, we present analysis results to answer our 3 research questions:

First we investigate the first research question - "Are there emotional differences between two types of tweets specifically related to the incident and not specifically related to the event?" and the third research question "What type of negative emotions are presented more during the terrorist attack?"

In order to answer the above two research questions, we compared consolidated means of the three expressed emotions during the 5 days, for both EB and nonEB groups; we then ran Analysis of Variance (ANOVA). Table 2 shows ANOVA results between two groups (EB and non-EB) for three negative emotions and the percentage mean value of each expressed emotion. The mean value of each emotion calculated by considering all of each emotion's expressed sentiment score during the incident.

Table 2. Analysis of variance result by group

	Consolidated mean	Mean of EB	Mean of non-EB	Mean square	F	p
Anger	1.465	1.610	1.320	4.226	2.449	0.119
Anxiety	0.536	0.474	0.597	20.321	12.26	0.001
Sadness	0.652	0.335	0.969	0.759	134.97	0.000

The consolidated mean shows that anger was the most highly expressed emotion during the 5 days. Consolidated mean of Anger was 1.465 % and this was around 2.5 times higher than any other emotion. This indicates that during the Boston bombing tragedy people expressed anger via the SNS more than Anxiety and Sadness.

This can be compared with other extreme event study cases, studied in prior literature, such as natural disaster (Japan earthquakes) [33] and another terror event (911 tragedy) [8]. In the case of natural disaster events, people expressed fear and anxiety on SNS rather than unpleasantness and anger. In addition, emotion expression regarding the 911 terrorist attacks showed that anger was the most strongly expressed emotion during the event period on the text pagers network. Our result confirmed that expressed emotions reflect the characteristic of the extreme event. For instance, terrorist attacks mostly made people angry, but on the other hand, natural disaster incidents made people worry about others and these emotions are clearly expressed on SNS.

Moreover, Table 2 shows whether the mean difference between the EB group and the non-EB group was statistically significant for three negative emotions (Anger, Anxiety and Sadness). The result indicates that for Anxiety and Sadness there are statistically significant differences in the expressed emotion between EB and non-EB groups (F = 12.26, p < 0.01, F = 134.97, p < 0.001). In addition, both Anxiety and Sadness were expressed higher (0.597 % and 0.969 %) when people did not include E&B keywords, which were directly related to the Boston bombing event. This implies people might avoid expressing emotions like anxiety and sadness in direct relation to extreme event related keywords.

Results for anger show that there is no statistically significant difference between the EB group and the non-EB group (F = 2.449, p > 0.05). This indicates that the emotion, anger, might not be impacted by the people's use of keywords. We interpreted this result as being derived from the characteristics of the extreme event. Since the Boston Bombing tragedy had infuriated people, the enhanced level of this anger could not be controlled, and this is what the SNS users expressed on SNS. An implication is that, if the disaster response authorities wish to mitigate people's expressed emotion, coping strategies needs to be separated between Anger and the other 2 emotions.

Table 3. shows ANOVA results by the day for three negative emotions, as well as the percentage mean value of each expressed emotion. The result indicates that for all three emotions (Anger, Anxeity, Sadness), there are statistically significant expressed emotion differences during at least one of the five days (F = 5.706, p < .001, F = 3.002, p < 0.05, 11.698, p < 0.001).

Table 3. Analysis of variance result by days

	Consolidated mean	Mean of day1	Mean of day2	Mean of day3	Mean of day4	Mean of day5	Mean square	F	p
Anger	1.401	1.059	1.752	1.990	1.327	0.878	9.608	5.706	0.000
Anxiety	0.532	0.510	0.560	0.613	0.537	0.441	0.189	3.002	0.020
Sadness	0.655	0.665	0.947	0.714	0.628	0.320	2.420	11.698	0.000

In order to investigate the impact of announcements or news on expressed emotions on SNS, we performed content analysis using hourly data collected during the 5 days of the Boston marathon bombing related tweet messages. By mapping the important event related announcements or news to the expressed sentiment change we can perhaps answer the remaining two research questions - "Do incident related announcements or news have an impact on people's emotions?" and "How long does it take those announcements or news to impact actual expressed emotions?"

4.1 Day 1

According to the content analysis results from both Groups of EB tweet data set and non-EB tweet data set, the major event for day 1 was the announcement that there were explosions in front of the Boston marathon finish line. Boston.com reported this as "2: 49 pm: There is an explosion in front of Marathon Sports on Boylston Street, close to the Marathon finish line. Thirteen seconds later and a block away, there is a second explosion in front of the Forum restaurant. Three people are killed, 282 are injured."

EB sentiment analysis result for day 1 showed peaks for all three emotions: anger, anxiety and sadness 4–5 h after the event happened. The change in emotion for anger, anxiety and sadness increased by 500 %, 112 % and 330 % respectively around 4 h after the incident. This indicates that the news about the bombing tragedy affected the change in all three emotions 4 h after the actual event.

Non-EB results also showed similar result for all three emotions. These three emotions significantly changed compared to the emotions expressed in the previous hours. The level of anger, anxiety, and sadness increased by 45.57 %, 44 %, and 77.05 % respectively also around 4–5 h after the actual event. The result of day 1 for both groups indicates that around 4–5 h later people's emotions were expressed on SNS and the degree of emotion change on keyword usage was much larger than that of non-keyword use group.

4.2   Day 2

For both the EB group and the non-EB group, we selected the announcement that "bombs were made by packing explosives, shrapnel and nails into pressure cookers. They are still unaware as to who detonated them and why." as a major event. This announcement started from 6 pm (GMT) and one emotion, anger, showed significant emotion change during the time period between 6 pm (GMT) and 8 pm (GMT). EB results showed that emotion change for anger showed both peaks and troughs in succession. Right after the announcement, a delta value of anger increased by 89.75 % then suddenly plunged to −85.11 % in the next hour. We assumed that the announced unclear information such as "unawareness" of reason and "suspicion" impacted people's emotions. For the non-EB group around 5–6 h after the announcement, 10 pm (GMT), there was a rise in the emotions of anxiety (52.83 %) and sadness (74.31 %). We interpret that levels in anxiety change occurred as a result of the broadcast about unawareness of the reason and suspects. However, we do not have an appropriate reason to explain the change in sadness.

4.3   Day 3

On day 3 a major announcement about the incident was "Several media outlets, starting with the CNN, report that an arrest has been made; BPD and FBI deny the report". Although the level of expressed emotion for anxiety and sadness were very low, EB results showed continuous fluctuation in all three emotions. Moreover, the non-EB results also reveal similar patterns regarding fluctuation of anxiety and sadness during the 4 h time period from 6 pm (GMT). This indicates that the diffused unverified information impacted people's negative emotions. However, when people did not include directly related keywords (E&B) in the message, anger did not have large fluctuations like the other emotions.

4.4   Day 4

Around 9 pm (GMT), FBI published surveillance photos of the bombing suspects and media around the world started broadcasting this announcement to the public. From both EB and non-EB results, we observed decreased negative emotions. The EB result shows that 1–2 h after the announcement, the levels of anxiety (–79.17 %) and sadness (–86.36 %) dropped compared to the levels before the announcement. In addition, the non-EB result showed that within 2 h of the announcement all three negative emotions anger (–46.63 %), anxiety (–44.44 %), and sadness (–67.03 %) decreased compared to before the announcement. We surmised that by making the announcement, FBI had reduced the uncertainty about suspects.

4.5 Day 5

The early morning 7 am (GMT) announcement about the investigation situation and the alert "Boston, Watertown has issued a 'shelter in place' advisory asking residents to stay in their homes as police continue their search for Tsarnaev. All mass transit is shut down" was also selected as a major event on day 5. Our sentiment result from EB groups showed that right after the announcement, people's anger, anxiety and sadness decreased –84.46 %, –87.75 % and –81.25 % but increased drastically in the following hour 86.96 %, 166.67 % and 600 % respectively. This indicates that when people heard about the situation, their level of negative emotions decreased but again increased because of the absence of any follow-up announcements. Moreover, the non-EB group presented opposite patterns compared with the EB group. Specifically, increased negative emotion were revealed within the first hour (anger 94.34 % and sadness 463.64 %) then people showed decreased levels of emotions in the following hours (anger –80.58 %, anxiety (–50 %), and sadness –6.45 % (1st hour) 65.52 % (2nd hour)). This implies that emotions among users who were including direct event related keywords and those who did not include keywords can be different. Therefore, when disaster response authorities deal with two different groups of SNS user, they need to consider heterogeneity between the two groups of users.

Figure 1 below shows an example of presented emotion change and the impact of the announcement for both EB and non-EB groups on Day 2. Graphs for all other days are in the appendix (Fig. 2).

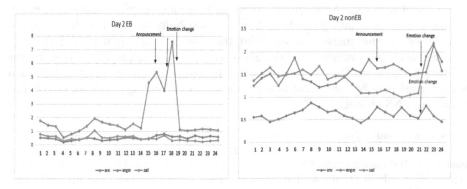

Fig. 1. Impact of announcement on emotion change

5 Discussion and Conclusion

In this research, we investigated three research questions "Are there emotion differences between two types of tweets", "what types of negative emotions are presented more during the terrorist attack", "what is the impact of an announcement on expressed emotion". Our analysis shows that for anxiety and sadness, differences in expressed emotion existed between the EB and non-EB group. In addition, anger was the most highly expressed emotion during the Boston bombing tragedy. By mapping our content analysis results into hourly-analyzed sentiment analysis results, we could see the impact of announcements on expressed emotion via SNS. Moreover, emotion change usually happened between 1 and 5 h from the announcement.

This study made the following theoretical and practical contributions. The study showed that expressed emotions were impacted by announcements/news, which had an effect on other user's emotions. In addition, based on the concept of warning signal keywords, we could separate the dataset into two groups called event related and non-event related groups and we could investigate the emotional differences among these two groups. The practical contribution of this study is that we can provide various information regarding expressed emotions during extreme events to disaster response authorities. Our analysis result indicates that in order to mitigate expressed negative emotion on SNS, disaster response agencies need to have specified strategies depending on the types of negative emotions and public keyword usage. In addition, depending on the extreme event context, mainly expressed type of negative emotion is varied. Therefore, developing mitigation strategies based on the characteristic of the extreme event is essential.

We also saw the impact of announcements from broadcasting system or government agencies on people's expressed emotions. For example, unverified announcements or lack of updates led to fluctuations of emotions. Finally, 1–5 h from announcement was the common interval between reactions and the actual announcement time. Therefore, if disaster response agencies want to mitigate the public's negative emotion, swiftness of action must also be considered.

There are however, some limitations to this study. First of all, we have only considered the Boston bombing tragedy as a case of extreme event. This may limit the generalizability of our findings. Second, we used a published timeline for content analysis and mapping. Although, the news in timelines described the major events during the Boston bombing incident, there is a possibility that we may have skipped other major announcements or news. Third, there are other widely used SNS channels such as Facebook, Instagram, or Google + but in this research; we focused on the Twitter SNS channel. Thus, as a future research, comparing the results of each popular SNS channel may be considered to improve our understanding regarding the characteristics of SNS channels.

Acknowledgements. This research is funded by the National Science Foundation (NSF) under grants 1353119 and 1353195. This research has also been funded in part by the National Science Foundation (NSF) under Grants 1241709, 1227353, 1419856 and 1554373. The usual disclaimer applies. We would like to thank Chulwhan Chris Bang for his data collection help and Chandrakanth Saravanan, Megan Saldanha and Swati Upadhya for research support.

Appendix

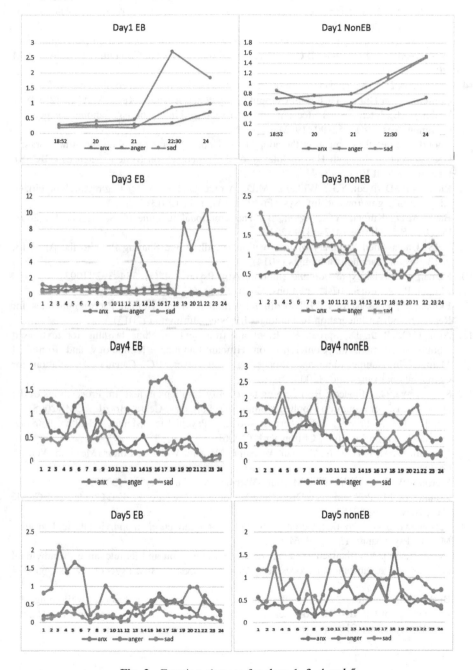

Fig. 2. Emotion changes for days 1, 3, 4 and 5

References

1. Pacherie, É.: Emotion and action. Eur. Rev. Philos. **5**, 55–90 (2002)
2. Hancock, J.T., et al.: I'm sad you're sad: emotional contagion in CMC. In: Proceedings of the 2008 ACM Conference on Computer Supported Cooperative Work, San Diego, CA, USA. ACM (2008)
3. Hancock, J.T., Landrigan, C., Silver, C.: Expressing emotion in text-based communication. In: Proceedings of the SIGCHI Conference on Human Factors in Computing Systems. ACM (2007)
4. Kramer, A.D.: The spread of emotion via Facebook. In: Proceedings of the SIGCHI Conference on Human Factors in Computing Systems, Austin, TX, USA. ACM (2012)
5. Walther, J.B.: Interpersonal effects in computer-mediated interaction a relational perspective. Commun. Res. **19**(1), 52–90 (1992)
6. Choudrie, J., et al.: Comparing the adopters and non-adopters of online social networks: a UK perspective. In: 46th Hawaii International Conference on System Sciences (HICSS), Hawaii, USA. IEEE (2013)
7. Rana, N.P., Dwivedi, Y.K., Williams, M.D.: A meta-analysis of existing research on citizen adoption of e-government. Inf. Syst. Front. **17**(3), 1–17 (2013)
8. Back, M.D., Küfner, A.C., Egloff, B.: The emotional timeline of September 11, 2001. Psychol. Sci. **21**(10), 1417–1419 (2010)
9. DHS, Unclassified Summary of Information Handling and Sharing Prior to the April 15, 2013 Boston Marathon Bombings (2014)
10. Ekman, P.: Facial expression and emotion. Am. Psychol. **48**(4), 384–392 (1993)
11. Ekman, P.: Argument basic emotions. Cogn. Emot. **6**(3–4), 169–200 (1992)
12. Ekman, P., Friesen, W.V., Ellsworth, P.: Emotion in the Human Face: Guidelines for Research and An Integration of Findings. Elsevier, Burlington (2013)
13. Alm, C.O., Roth, D., Sproat, R.: Emotions from text: machine learning for text-based emotion prediction. In: Conference on Human Language Technology and Empirical Methods in Natural Language Processing, Vancouver, BC, Canada. Association for Computational Linguistics (2005)
14. Strapparava, C., Mihalcea, R.: Learning to identify emotions in text. In: Proceedings of the 2008 ACM Symposium on Applied Computing, Fortaleza, Ceara, Brazil. ACM (2008)
15. Frisch, N.C., Frisch, L.E., Frisch, L.E.: Psychiatric Mental Health Nursing. Delmar/Thomson Learning, Clifton Park (2006)
16. Merriam-Webster, Fear, in Merriam-Webster, Merriam-Webster, Editor. Merriam-Webster (2015)
17. Merriam-Webster, Sadness, in Merriam-Webster. Merriam-Webster (2015)
18. Konnikova, M.: Problems Too Disgusting to Solve, in The New Yorker. Condé Nast: Condé Nast, New York (2015)
19. Schoenewolf, G.: Emotional contagion: behavioral induction in individuals and groups. Modern Psychoanal. **15**(1), 49–61 (1990)
20. Vijayalakshmi, V., Bhattacharyya, S.: Emotional contagion and its relevance to individual behavior and organizational processes: a position paper. J. Bus. Psychol. **27**(3), 363–374 (2012)
21. Burke, M., Marlow, C., Lento, T.: Social network activity and social well-being. In: Proceedings of the SIGCHI Conference on Human Factors in Computing Systems, Atlanta, GA, USA. ACM (2010)

22. Ma, Q., Jin, J., Wang, L.: The neural process of hazard perception and evaluation for warning signal words: evidence from event-related potentials. Neurosci. Lett. **483**(3), 206–210 (2010)
23. Correll, J., Urland, G.R., Ito, T.A.: Event-related potentials and the decision to shoot: the role of threat perception and cognitive control. J. Exp. Soc. Psychol. **42**(1), 120–128 (2006)
24. Qin, J., Han, S.: Neurocognitive mechanisms underlying identification of environmental risks. Neuropsychologia **47**(2), 397–405 (2009)
25. Nasukawa, T., Yi, J.: Sentiment analysis: capturing favorability using natural language processing. In: Proceedings of the 2nd International Conference on Knowledge Capture, Sanibel Island, FL, USA. ACM (2003)
26. Stieglitz, S., Krüger, N.: Analysis of sentiments in corporate Twitter communication–a case study on an issue of Toyota. Analysis **1**, 1–2011 (2011)
27. Pennebaker, J.W., et al.: The development and psychometric properties of LIWC 2007 (2007)
28. Tausczik, Y.R., Pennebaker, J.W.: The psychological meaning of words: LIWC and computerized text analysis methods. J. Lang. Soc. Psychol. **29**(1), 24–54 (2010)
29. Tumasjan, A., et al.: Election forecasts with Twitter: how 140 characters reflect the political landscape. Soc. Sci. Comput. Rev. **29**(4), 402–418 (2010)
30. Yu, B., Kaufmann, S., Diermeier, D.: Exploring the characteristics of opinion expressions for political opinion classification. In: Proceedings of the 2008 International Conference on Digital Government Research. Digital Government Society of North America (2008)
31. O'Neill, A.: Tsarnaev trial. In: Timeline of the bombings, manhunt and aftermath. CNN: CNN (2015)
32. Morrison, S., O'Leary, E.: Timeline of Boston marathon bombing events. In: Boston.com., Boston.com: Boston.com (2015)
33. Vo, B., Collier, N.: Twitter emotion analysis in earthquake situations. Int. J. Comput. Linguist. Appl. **4**(1), 159–173 (2013)

Discovering the Voice from Travelers:
A Sentiment Analysis for Online Reviews

Wei-Lun Chang[✉]

Department of Business Administration,
Tamkang University, New Taipei City, Taiwan
wlchang@mail.tku.edu.tw

Abstract. We proposed a model on the basis of a sentiment analysis in terms of positive and negative words and the concept of credibility inferred from prospect theory. This study used TripAdvisor to examine the proposed model and selected 10 out of 271 hotels in Las Vegas between January and February 2015, which is the peak season for traveling to Las Vegas. Through the sentiment analysis, we determined that the overall ranking of the 10 hotels decreased. However, the ranking of Skylofts at MGM Grand decreased by only 1 level. The rankings of the 9 other hotels decreased by 2 or more levels after the credibility factor was considered. Moreover, the credibility factor affected the overall ranking. Apparently, the credibility factor has a higher influence on hotel ranking than the sentiment analysis does. These results revealed that negative emotions and low-credibility reviews have a high influence on hotel ranking.

Keywords: Sentiment analysis · Prospect theory · Online review comments

1 Introduction

The popularity of the Internet enables people to change lifestyle through various means, such as sharing and searching information [1]. Gretzel, Yoo, and Purifoy [2] revealed that 96.4 % of people search online information before booking hotels. In 2013, PhoCusWright conducted a research on the users of TripAdvisor, and the results revealed that more than half of the users did not make decisions until they had read past reviews from visitors. In addition, certain researchers have insisted that overall ratings represent the perceptions toward products [3, 4]. However, others have reported that overall ratings cannot represent the emotions of customers, particularly using positive and negative emotional words, customer preference ranking, and the evaluation of product features to replace overall ratings [5–7]. Thus, overall rating is an incomplete mechanism for evaluating products.

The anonymity of virtual platforms may result in fake reviews [9]. Chesney [10] questioned the credibility of Wikipedia and designed an experiment to evaluate its reputation. Racherla [11] discovered that false information results in user distrust in an online platform. The results also revealed that the review content and customer perceptions of the reviews are the major factors that influence trustworthiness. Several researchers have questioned the credibility of overall ratings and identified the significance of trust in reviews. However, existing studies have used only a five-star rating

© Springer International Publishing Switzerland 2016
V. Sugumaran et al. (Eds.): WEB 2015, LNBIP 258, pp. 15–26, 2016.
DOI: 10.1007/978-3-319-45408-5_2

system to represent overall ratings and the lack of trust. Users do not read all review content when making decisions because of the large amount of online information.

BrightLocal [12] reported that 67 % of online users in North America read fewer than six reviews before purchase. Short [13] investigated passengers using online reviews and observed that 40 % and 29 % of them agreed that concrete reviews and overall ratings influence decisions, respectively. In addition, 72 % out of 90 % users who read online reviews trust companies because of positive reviews. Thus, emotions expressed in reviews are crucial during the decision-making process. O'Mahony and Smyth [7] also used positive and negative terms to evaluate the usefulness of reviews. Koh [8] specified that review content, overall ratings, and emotions influence the online decision-making process. The results revealed that users rank overall ratings to search information, filter data, analyze emotions, and make decisions. Patel [14] reported that emotion is the most crucial factor that influences online sales. In addition, we observed that online reviews are diverse, and reading every review wastes time.

Researchers have indicated that, in addition to overall ratings, useful reviews are meaningful for reference to an online platform. Long et al. [6] identified that efficient overall ratings are meaningful for reference to consumers for decision making. Existing studies have considered only the credibility of reviews, not emotions. Hence, this study proposed two research questions: (a) What are the key factors of reviews and how can a quantitative review model be constructed? and (b) How crucial are emotions in the review model? This study reviewed the literature to emphasize the importance of emotions and used the concept proposed by Koh [8] to quantify emotions on the basis of the case of TripAdvisor. Moreover, Gretzel, Yoo, and Purifoy [2] indicated that credibility should be measured on the basis of the reputation of reviewers. The results showed that 75.3 % of people consider that the travel experience of reviewers is crucial, 65.9 % of people compare the similarity of travel experiences between themselves and reviewers, and 58.5 % of people judge credibility on the basis of the review content. Thus, this study considered the concept of credibility in the proposed review model. The research goals were as follows: (a) constructing a sentiment- and credibility-based review model (e.g., quantifying emotions in review content) and (b) emphasizing the importance of the credibility of reviews (e.g., considering the travel experiences of reviewers).

2 The Proposed Model

This study used an online review platform, which includes overall ratings and review content, as the basis for the proposed model. The researchers indicated that overall rating cannot satisfy user needs and attempted to improve recommendation mechanisms [5–7]. In particular, O'Mahony and Smyth [7] specified combining reviewer reputation, degree of interaction, and sentiment terms for finding useful review comments. Thus, reviewer and review content were two major components in our model. Certain researchers have also identified the importance of sentiment in word of mouth [15–17]. Credibility is a key factor for review content [11, 18–20]. This study divided reviews into two major parts: reviewer and review content. The sentiment factor was used for review content, and credibility was used for reviewers. In Eq. (1), *Sentiment* is the value from review content based on the sentiment analysis, and *credibility* is the function for adjusting the sentiment value. The adjusted score is represented as *Review_Score*, which is the score of an individual review.

$$Review_{score} = f_{credibility}(Sentiment) \tag{1}$$

According to the literature review, risk is a key factor affecting online shopping [21–25]. In addition, several researchers have specified that trust is the crucial factor affecting online shopping [25, 26]. Zimmer [27] argued that consumers with high trust may have low risk perception. Consequently, this study considered that risk and credibility are related and that both influence online shopping decisions.

In addition, Racherla [11] proposed that consumer behavior is influenced by psychological factors. Online platforms are also anonymous; hence, they may result in fake reviews [9]. Thus, this study considered that reading reviews existed risk. The proposed model is based on prospect theory, which was proposed by Kahneman and Tversky [28]. This theory indicates that people have different attitudes toward gain and loss, and individual perceived bias may result in decision differences.

Prospect theory explains that people exhibit risk seeking when facing losses and risk aversion when facing gains; in other words, perceived loss is more sensitive than gain. In addition, the reference point for separating risk seeking from risk aversion is different for different people. This study considered that users may perceive different levels of gain when reading online reviews because they do not incur loss in this situation. In addition, according to prospect theory, we assumed that users have no preference (neutral); that is, no fixed reference point was used in our model. If review content is highly trusted, perceived risk decreases. If review content is less trusted, perceived risk increases. Thus, risk seeing slightly affects the estimated scores of online reviews, and risk aversion affects them dramatically. It is assumed that if the reviewers are more experienced or certified by third parties in an uncertain environment (e.g., the Internet), reviews are more credible. Gretzel et al. [2] and Bianchi and Andews [29] confirmed that reviewer reputation is linked to the credibility of reviews when the reviewers are experienced. Hence, emotion is the first factor for estimating the review score, and credibility is the second factor for adjusting the score accordingly in our model. Moreover, we assumed that risk exists on online platforms, and the literature has revealed that highly reputed reviewers represent high credibility of reviews. Thus, the risk of trusting an online review is low if the reviewer is highly credible.

2.1 Sentiment Factor

Studies have analyzed the influence of word of mouth on sales [3, 4, 30–39]. However, these studies investigated only the ranking and overall rating mechanism of reviews and neglected the importance of review content. Users seek useful information from reviews and not from overall ratings [3], and sentiments critically influence purchase decisions [15–17]. In addition, certain researchers have considered that the overall rating of a review cannot present real emotions [5–7]. Thus, this research includes review content in the quantitative sentiment analysis to complement overall rating.

O'Mahony and Smyth [7] measured the usefulness of review content in terms of positive and negative emotional words and specified using positive emotional words to represent the usefulness of a review. Moreover, Koh [8] discovered that overall rating might be different from the review content. For example, the three-star overall rating might be inconsistent with the described content that includes certain positive feedback. Thus, review content (text) can deliver more useful messages in the sentiment analysis. This study revised the concept of Koh [8], as shown in Eq. (2).

$$Sentiment_i = \frac{(Senti\_Score_{title} + Senti\_Score_{content})}{2} \tag{2}$$

In Eq. (2), $Sentiment_i$ represents the sentiment score of each review, $Senti\_Score_{title}$ is the sentiment score of the review title for a specific review topic i, and $Senti\_Score_{content}$ is the sentiment score of review content for a specific review topic i. The concept is to calculate the average of sentiment scores of the title and content. Both review title and content are considered because previous researchers mostly analyzed only the review content for sentiments. In addition, numerous websites, such as Amazon, Rotten Tomatoes, TripAdvisor, and Yelp, provide two major parts (title and content). The survey of BrightLocal [12] revealed that 67 % of consumers in North America read no more than six reviews before making decisions. Consequently, this study considered that the sentiment construct should include both review title and content.

Equation (3) represents the equation for estimating sentiment scores. Koh [8] also emphasized that using only positive and negative emotional words or strong positive and negative emotional words is not holistic. Thus, the model in Eq. (3) includes strong positive (str_pos_i), strong negative (str_neg_i), ordinary positive (ord_pos_i), and ordinary negative (ord_neg_i) emotional words, and the weight (wg), indicates the sum of all positive and negative emotional words, indicates the difference between all positive and negative emotional words. To identify the weight of different emotional terms, this study assigned +2 for str_pos_i and str_neg_i and +1 for ord_pos_i and ord_neg_i. The concept is to estimate the ratio of positive emotional words to the total emotional words.

$$Senti\_Score_i = \frac{(str\_pos_i * wg_1 + ord\_pos_i * wg_2) - (str\_neg_i * wg_1 + ord\_neg_i * wg_2)}{(str\_pos_i + str\_neg_i) * wg_1 + (ord\_pos_i + ord\_neg_i) * wg_2} \tag{3}$$

According to Eq. (3), the range of sentiment scores was [−1,1]. In addition, we observed that most online review platforms used a five-star rating system to represent overall rating (1 for *extremely bad*, and 5 for *extremely good*). Hence, this research converts the estimated sentiment score (x) into the traditional presentation format (e.g., 1 to 5), as shown in Eq. (4). After conversion, the final score is 3 if the estimated sentiment score is 0, 1 if the estimated sentiment score is −1, and 5 if the estimated sentiment score is 1.

$$f'(Sentiment_i) = 2 * Sentiment_i + 3 \tag{4}$$

This study used an online dictionary from thesaurus.com as the base to discover emphasized positive and negative emotional words (Appendix). The emphasized adverbs included words such as *very* and *really*, the positive emotional adjectives included words such as *kind* and *nice*, and the negative adjectives included words such as *sad* and *terrible*. If a review has the phrase *very clean*, which includes emphasized and positive terms, the sentiment score is +2 (i.e., the weight is +2, and the number of vocabulary is 1). If a review has only the term *clean*, the sentiment score is 1 (i.e., the weight is +1, and the number of vocabulary is 1).

2.2 Credibility Factor

In addition to the sentiment analysis of reviews, this study proposed a novel concept of credibility for adjusting the estimated scores. We classified reviewers into five levels based on the number of reviews. This is the most common approach used by websites such as TripAdvisor. For example, *reviewer* is used for reviewers with 3–5 reviews (level 1), *senior reviewer* for 6–10 reviews (level 2), *contributor* for 11–20 reviews (level 3), *senior contributor* for 21–49 reviews (level 4), and *top contributor* for ≥ 50 reviews (level 5). In this study, we used 80/20 rules to segregate the five levels of reviewers into two categories. Consequently, level 3, 4, and 5 reviewers belong to the more trustworthy category, and level 1 and 2 reviewers belong to the less trustworthy category.

$$Review\_Score = y = \log_a f'(Sentiment_i), \text{ if } a > 1 \tag{5}$$

In Eq. (5), indicates the converted sentiment score, with values between 1 and 5, and y stands for the adjusted sentiment score with high credibility. This study used prospect theory for the perceived gain perspective in Eq. (5). The concave curve of high credibility in Fig. 1 indicates that the sentiment score is adjusted slightly if the reviewers are trustworthy (i.e., levels 3, 4, and 5). High-level reviewers may provide less risk of review content on the Internet, which is an uncertainty environment. The adjusted sentiment score should be approximately equal the original score when people do not fully trust online content.

$$Review\_Score = |y| = \log_b f'(Sentiment_i), \text{ if } 0 < b < 1 \tag{6}$$

In Eq. (6), the concept is to use the loss perspective of prospect theory. The variable indicates the converted sentiment score, which is between 1 and 5, and y stands for the adjusted sentiment score with less credibility. The convex curve of low credibility in Fig. 2 indicates that the sentiment score is adjusted dramatically if the reviewers are less trustworthy (i.e., levels 1 and 2). The converted sentiment score may have less influence on the user.

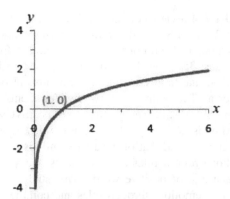

Fig. 1. The curve of high credibility

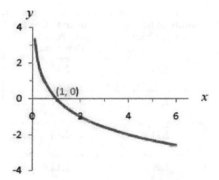

Fig. 2. The curve of low credibility

3 Data Analysis

3.1 Data Collection

This study used TripAdvisor as the platform for collecting data. According to the ranking from TripAdvisor, the top 10 hotels were selected out of 271 hotels in Las Vegas. We ruled out bias because English is the official language on TripAdvisor for providing comments. In addition, we collected review comments between January and February 2015 because New Year vacation is the peak season for travel. The order of the 10 hotels based on their original ranking from 1 to 10 was Mandarin Oriental, ARIA Sky Suites, Four Seasons Hotel, Skylofts at MGM Grand, Palazzo Resort Hotel, Wynn Las Vegas, Encore at Wynn, Staybridge Suites, Hotel32 at Monte Carlo, and Venetian Resort Hotel. We reranked the hotels on the basis of ratings during the data collection period. The new ranking was Skylofts at MGM Grand (overall rating 5), ARIA Sky Suites (4.86), Staybridge Suites (4.86), Hotel32 at Monte Carlo (4.82), Mandarin Oriental (4.69), Four Seasons Hotel (4.63), Encore at Wynn (4.63), Wynn Las Vegas (4.56), Palazzo Resort Hotel (4.55), and Venetian Resort Hotel (4.53).

This study collected 1,724 review comments; 7 from Skylofts at MGM Grand, 59 from ARIA Sky Suites, 21 from Staybridge Suites, 28 from Hotel32 at Monte Carlo, 137 from Mandarin Oriental, 43 from Four Seasons Hotel, 225 from Encore at Wynn, 324 from Wynn Las Vegas, 357 from Palazzo Resort Hotel, and 523 from Venetian Resort Hotel. We observed that hotels having high rankings also had positive review comments and vice versa. The number of review comments did not necessarily influence the ranking. In Table 1, the average number of positive words is at least 0.29, and that of negative words is at most 0.07 for review titles. The average number of positive words is at least 6.18, and that of negative words is at most 1.16 for review comments. For all the hotel review titles and comments, the average number of negative words is lower than that of positive words. This study inferred that customers intended to deliver positive emotions through titles and comments.

Table 1. Comparison between adjusted ratings based on positive and negative comments

Hotel	Original Rating	Adjusted Rating by Credibility	Adjusted Rating by only Positive Comments	Adjusted Rating by only Negative Comments
Skylofts at MGM Grand	5	2.59	2.82	1.20
ARIA Sky Suites	4.86	2.53	2.52	2.46
Staybridge Suites	4.86	2.33	2.33	0
Hotel32 at Monte Carlo	4.82	2.04	2.12	0.98
Mandarin Oriental	4.69	2.44	2.53	1.76
Four Seasons Hotel	4.63	3.10	3.12	2.88
Encore At Wynn	4.63	2.42	2.50	1.61
Wynn Las Vegas	4.56	2.50	2.56	1.62
The Palazzo Resort Hotel	4.55	2.43	2.52	1.61
Venetian Resort Hotel	4.53	2.50	2.60	1.74

Reviewers are the main source of reviews for all hotels; for example, 29 % for Skylofts at MGM Grand, 48 % for Staybridge Suites, 64 % for Hotel32 at Monte Carlo, 40 % for Mandarin Oriental, 19 % for Four Seasons Hotel, 40 % for Encore at Wynn, 41 % for Wynn Las Vegas, 45 % for Palazzo Resort Hotel, and 37 % for Venetian Resort Hotel. Thus, hotels may hire writers to write comments. The credibility of reviewers can be used to reduce false review comments. This study included both sentiment analysis and reviewer credibility in the proposed model to reconsider the ranking of hotels.

3.2   Sentiment Analysis of Review Titles and Comments

This study used LIWC (Linguistic Inquiry and Word Count) software to analyze the 1,724 review titles and comments. We used the concept proposed by Koh [8] to measure sentiment scores of the review titles and comments, which are represented as (*i* represents a specific review). Before estimating the sentiment scores, the number of positive and negative words in the titles and content should be identified. This study also segregated sentiment words into four categories: strongly positive words, strongly negative words, common positive words, and common negative words. Each category is given a different score; for example, strongly positive and negative words are assigned a score of 2, and common positive and negative words are assigned a score of 1. The final sentiment score of each review is calculated by averaging the sentiment score of the review title and comments. Regarding the review titles, ARIA Sky Suites had 0.07 strongly positive words on average, Encore at Wynn had 0.01 strongly negative words on average, Staybridge Suites had 0.76 common positive words on average, and Hotel32 at Monte Carlo and Four Seasons Hotel both had 0.07 common negative words on average. Regarding the review comments, Staybridge Suites had 1.81 strongly positive words on average, ARIA Sky Suites had 0.17 strongly negative words on average, Staybridge Suites had 6.81 common positive words on average, and Mandarin Oriental had 1.01 common negative words on average.

Result reveals that the sentiment scores of the review title and comments for Staybridge Suites were 0.62 and 0.94, respectively, which were the highest among all the hotels. This study indicated that a high number of positive words (strongly and common) resulted in high sentiment scores and vice versa. The results also revealed that people expressed emotions through review titles. This study considered review titles and comments simultaneously, which is more comprehensive than previous related studies. Moreover, the estimated sentiment score ranged from −1 to 1. We adjusted the rating range between 1 and 5 by using a converting function.

3.3   Analysis of Reviewer Credibility

After considering sentiments in the review titles and comments, this study applied the concept of credibility to each reviewer. Five levels of reviewers were distinguished and given different credits (i.e., reviewer, senior reviewer, contributor, senior contributor, and top contributor). This study employed two combinations of the log function. The first log function assumed high credibility of reviews written by contributors, senior contributors, and top contributors. The second log function assumed low credibility of reviews written by reviewers and senior reviewers. Two parameters, a and b, were assigned to the two log functions. This study determined a as 1.5 and b as 0.4 after trial and error to obtain the real values. The logic was to clearly differentiate between the high and low credibility of reviewers. Result reveals that the adjusted ratings based on credibility of Hotel32 at Monte Carlo and Four Seasons Hotel were 2.04 (lowest) and 3.1 (highest), respectively. We inferred that this is because most reviewers of Hotel32 at Monte Carlo had low credibility (i.e., reviewers and senior reviewers), whereas those of Four Seasons Hotel had high credibility (i.e., contributors, senior contributors, and top contributor). The average numbers of reviews of each reviewer for Four Seasons Hotel and Hotel32 at Monte Carlo were 28 (highest) and 8.11 (lowest), respectively. This study revealed that more reviews generated higher adjusted ratings, and fewer reviews generated lower adjusted ratings.

The finding reveals that Venetian Resort Hotel had the maximum number of reviewers (195) and top contributors (61). Venetian Resort Hotel is a famous hotel on the Las Vegas Strip. Moreover, Palazzo Resort Hotel is adjacent to Venetian Resort Hotel and belongs to the same group. These hotels have a large number of travelers and comments because of people who love similar types of hotels. A similar situation was observed for Wynn Las Vegas and Encore at Wynn. Four Seasons Hotel had 9 contributors, 11 senior contributors, and 13 top contributors, which was much higher than the number of reviewers (8) and senior reviewers (2). We inferred that travelers had extensive experience of staying at Four Seasons Hotel, which placed the final rating at the top position, although the total number of reviews was not high. Similarly, Hotel32 at Monte Carlo had 18 reviewers, which was much higher than the others. Thus, the adjusted rating of Four Seasons Hotel increased because of the large number of higher-credibility reviewers. The adjusted rating of Hotel32 at Monte Carlo decreased because of the large number of lower-credibility reviewers. The results also proved the importance of the credibility of reviewers in the proposed model.

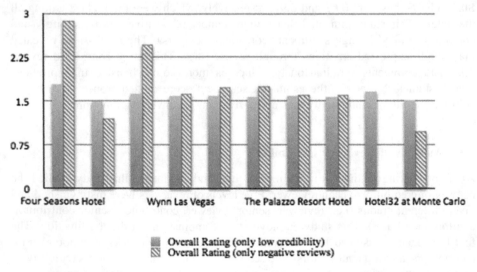

Overall Rating (only low credibility)
Overall Rating (only negative reviews)

Fig. 3. Adjusted ranking of low-credibility reviewers and negative comments

4 Conclusion

The results revealed that the adjusted ratings of Four Seasons Hotel were higher than those of the other hotels, irrespective of whether only positive (3.12) or negative comments (2.88) were considered (Table 1). However, although the adjusted rating of Hotel32 at Monte Carlo based on only positive comments was the lowest (2.12), that based on only negative comments was not the lowest. Thus, we inferred that negative comments substantially influenced the ratings compared with positive comments. Particularly, although Staybridge Suites had no negative comments, its adjusted rating was not the highest. This may have been because reviewers of different levels can affect the credibility of reviews.

We observed that high-credibility reviewers provided more negative comments than low-credibility reviewers did for Four Seasons Hotel. This was because high-credibility reviewers were more experienced and had high expectations regarding hotel service and facilities. Hence, we inferred that low-credibility reviewers were the key factor that influenced hotel rankings. ARIA Sky Suites and Staybridge Suites had the same original rating (4.86); however, the adjusted rating of ARIA Sky Suites (2.53) was higher than that of Staybridge Suites (2.33). Figure 3 shows the effect of negative reviews and low-credibility reviewers on our model. Result indicates that adjusted rankings did not substantially differ among all the hotels when considering only low-credibility reviewers (all ratings were approximately 1.5). However, the adjusted rankings differed significantly among all the hotels when considering only negative reviews (the highest rating was approximately 3 and the lowest was 0). For example, Four Seasons Hotel was at the top position even when only negative reviews were considered. We inferred that negative reviews substantially affected hotel rankings. Moreover, credibility was crucial to our model and may have affected the ranking even in the absence of negative reviews (e.g., Staybridge Suites).

5 Limitations

Although the proposed model has several advantages, it has limitations that can be improved by conducting additional studies. First, the segregation of reviewers was based on the mechanism used by TripAdvisor. Different platforms may have different mechanisms. In addition, we ignored the demographics of reviewers because of the difficulty in collecting data. If the platform owner can provide more detailed information on reviewers, detailed analysis may provide valuable insights. Second, we collected data only from TripAdvisor, which uses English as the official language for reviews. Finally, the sentiment analysis may not detect bias from review content such as irony. Although most review content truly reflects emotions, irony may be used occasionally. Such bias can be adjusted for to make the result more precise and valuable in the future.

References

1. Berger, J., Schwartz, E.M.: What drives immediate and ongoing word of mouth? J. Mark. Res. **48**(5), 869–880 (2011)
2. Gretzel, U., Yoo, K.H., Purifoy, M.: Online travel review study: role and impact of online travel reviews, Online Travel Review Report. Laboratory for Intelligent Systems in Tourism Texas A&M University (2007)
3. Chevalier, J.A., Mayzlin, D.: The effect of word of mouth on sales: online book reviews. J. Mark. Res. **43**(3), 345–354 (2006)
4. Clemons, E.K., Gao, G.G., Hitt, L.M.: When online reviews meet hyperdifferentiation: a study of the craft beer industry. J. Manag. Inf. Syst. **23**(2), 149–171 (2006)
5. Ganesan, K., Zhai, C.: Opinion-based entity ranking. Inf. Retrieval **15**(2), 116–150 (2012)
6. Long, C., Zhang, J., Huang, M., Zhu, X., Li, M., Ma, B.: Estimating feature ratings through an effective review selection approach. Knowl. Inf. Syst. **38**(2), 419–446 (2014)
7. O'Mahony, M.P., Smyth, B.: A classification-based review recommender. Knowl. Based Syst. **23**(4), 323–329 (2010)
8. Koh, N.S.: The valuation of user-generated content: a structural, stylistic and semantic analysis of online reviews, Ph.D. dissertation. Singapore Management University (2011)
9. Cochrane, K.: Why TripAdvisor is getting a bad review. The Guardian (2011). http://www.theguardian.com/travel/2011/jan/25/tripadvisor-duncan-bannatyne
10. Chesney, T.: An empirical examination of wikipedia's credibility. First Monday **11**(11) (2006). http://firstmonday.org/article/view/1413/1331
11. Racherla, P.: Factors influencing consumers' trust perceptions of online product reviews: a study of the tourism and hospitality online product review systems, Ph.D. dissertation. Temple University (2008)
12. Anderson, M.: Local Consumer Review Survey 2014. BrightLocal Blog (2014). http://www.brightlocal.com/2014/07/01/local-consumer-review-survey-2014/
13. Short, T.: Survey: How Travelers use Online Hotel Reviews. Overnight SUCCESS (2014). http://overnight-success.softwareadvice.com/survey-how-travelers-use-online-hotel-reviews-0614/
14. Patel, N.: How emotions affect the decision-making process. In: Online Sales. Forbes (2014) http://www.forbes.com/sites/neilpatel/2014/11/06/how-emotions-affect-the-decision-making-process-in-online-sales/
15. Ladhari, R.: The effect of consumption emotions on satisfaction and word-of-mouth communications. Psychol. Mark. **24**(12), 1085–1108 (2007)
16. Söderlund, M., Rosengren, S.: Receiving word-of-mouth from the service customer: an emotion-based effectiveness assessment. J. Retail. Consum. Serv. **14**(2), 123–136 (2007)
17. Sweeney, J.C., Soutar, G.N., Mazzarol, T.: The difference between positive and negative word-of-mouth—emotion as a differentiator. In: Paper presented at the Proceedings of the ANZMAC 2005 Conference: Broadening the Boundaries, pp. 331–337 (2005)
18. Hsiao, K., Lin, J.C., Wang, X., Lu, H., Yu, H.: Antecedents and consequences of trust in online product recommendations: an empirical study in social shopping. Online Inf. Rev. **34**(6), 935–953 (2010)
19. Mehrabi, D., Abu Hassan, M., Ali, S., Sham, M.: News media credibility of the internet and television. Eur. J. Sci. Res. **11**(1), 136–148 (2009)
20. Schlosser, A.E.: Source perceptions and the persuasiveness of internet word-of-mouth communication. Adv. Consum. Res. **32**(1), 202–203 (2005)
21. Bourlakis, M., Papagiannidis, S., Fox, H.: E-consumer behaviour: past, present and future trajectories of an evolving retail revolution. Int. J. E-Bus. Res. (IJEBR) **4**(3), 64–76 (2008)

22. Drennan, J., Sullivan, G., Previte, J.: Privacy, risk perception, and expert online behavior: an exploratory study of household end users. J. Organ. End User Comput. (JOEUC) **18**(1), 1–22 (2006)
23. Ha, H., Coghill, K.: Online shoppers in Australia: dealing with problems. Int. J. Consum. Stud. **32**(1), 5–17 (2008)
24. Kuhlmeier, D., Knight, G.: Antecedents to internet-based purchasing: a multinational study. Int. Mark. Rev. **22**(4), 460–473 (2005)
25. McCole, P., Ramsey, E., Williams, J.: Trust considerations on attitudes towards online purchasing: the moderating effect of privacy and security concerns. J. Bus. Res. **63**(9), 1018–1024 (2010)
26. Tan, F.B., Sutherland, P.: Online consumer trust: a multi-dimensional model. J. Electron. Commer. Organ. (JECO) **2**(3), 40–58 (2004)
27. Zimmer, J.C., Arsal, R.E., Al-Marzouq, M., Grover, V.: Investigating online information disclosure: effects of information relevance, trust and risk. Inf. Manag. **47**(2), 115–123 (2010)
28. Kahneman, D., Tversky, A.: Prospect theory: an analysis of decision under risk. Econometrica: J. Econometric Soc. **47**(2), 263–291 (1979)
29. Bianchi, C., Andrews, L.: Risk, trust, and consumer online purchasing behaviour: a chilean perspective. Int. Mark. Rev. **29**(3), 253–275 (2012)
30. Chintagunta, P.K., Gopinath, S., Venkataraman, S.: The effects of online user reviews on movie box office performance: accounting for sequential rollout and aggregation across local markets. Mark. Sci. **29**(5), 944–957 (2010)
31. Dellarocas, C., Awad, N., Zhang, X.: Exploring the value of online reviews to organizations: implications for revenue forecasting and planning. In: ICIS 2004 Proceedings, vol. 30 (2004)
32. Dellarocas, C., Narayan, R.: A statistical measure of a population's propensity to engage in post-purchase online word-of-mouth. Stat. Sci. **21**(2), 277–285 (2006)
33. Duan, W., Gu, B., Whinston, A.B.: Do online reviews matter?—An empirical investigation of panel data. Decis. Support Syst. **45**(4), 1007–1016 (2008)
34. Godes, D., Mayzlin, D.: Using online conversations to study word-of-mouth communication. Mark. Sci. **23**(4), 545–560 (2004)
35. Godes, D., Silva, J.: The dynamics of online opinion, Preliminary draft of unpublished manuscript. Harvard University (2006)
36. Hu, N., Pavlou, P.A., Zhang, J.: Can online reviews reveal a product's true quality?: Empirical findings and analytical modeling of online word-of-mouth communication. In: Paper presented at the Proceedings of the 7th ACM Conference on Electronic Commerce, pp. 324–330 (2006)
37. Liu, Y.: Word of mouth for movies: its dynamics and impact on box office revenue. J. Mark. **70**(3), 74–89 (2006)
38. Li, X., Hitt, L.M.: Self-selection and information role of online product reviews. Inf. Syst. Res. **19**(4), 456–474 (2008)
39. Moe, W., Trusov, M.: The value of social dynamics in online product ratings forums. J. Mark. Res. **48**(3), 444–456 (2011)

Design of Intelligent Agents
for Supply Chain Management

YoonSang Lee and Riyaz Sikora[✉]

Department of Information Systems, College of Business,
University of Texas at Arlington, P.O. Box 19437, Arlington, TX 76019, USA
yoonsang.lee@mavs.uta.edu, rsikora@uta.edu

Abstract. Using intelligent agents can be a good alternative for the automated supply chain management in e-commerce environment and decision support in current commerce practices [8]. This study focuses on finding the optimal structure of intelligent agents that yield the best performance for supply chain management. This study was conducted in two phases. In the first phase, a model for agent was developed and implemented. In the model we applied Q-learning, Softmax function, and ε-greedy to control the inventory threshold dynamically and used a sliding window protocol for flexible bidding strategy. Also, a testing environment with competing agents was implemented. In the second phase, two agents of different types were tested against each other in the same simulation. This simulation was played twice to compare our agent with two other types of agents. Results of simulations shows that our agent has better performance in two different simulations.

Keywords: Multi-agent system · Intelligent agent · Supply chain · Machine learning · Trading agent

1 Introduction

As the range of business expanded to a global level, many businesses have widened their scope of activities into the world wide, and this expanded market has provided the businesses more suppliers and customers than previous local-limited market. In addition to the trend of globalization, e-commerce became popular and provided additional options for companies' procurement and sales managers. With these two new trends, many companies have changed their strategy for supply chain management. Rather than manufacturing all of the parts in their factory, companies purchase most of the parts from various suppliers in the world, manufacture final products by assembling the purchased parts and the parts they made in-house, and sell the completed products to the customers in various countries.

According to the study by ACQUITY GROUP [1], 68 % of B2B buyers made online purchases in 2014 versus 57 % in 2013. Nearly half (46 %) of the buyers spent at least 50 % of their corporate procurement budgets online in the last year. However, a supply chain has many elements such as suppliers, customer, factory (production), shipment, and inventory, and these elements interact with each other. In the business practice, companies have to manage large number of dynamic factors created by

© Springer International Publishing Switzerland 2016
V. Sugumaran et al. (Eds.): WEB 2015, LNBIP 258, pp. 27–39, 2016.
DOI: 10.1007/978-3-319-45408-5_3

elements such as suppliers' supply price change, sudden increase of customer demand, limited manufacturing capacity, and so on. Because of the difficulties in managing a supply chain, supply chain is recognized as a complex adaptive system requiring dynamic and flexible reactions against the changes of environment for its effective management [7].

When considering the characteristics of supply chain, using an automated and adaptive agent can be a good alternative for the automation of supply chain management in an e-commerce environment and for decision support in current commerce practices [8]. While a theoretical model of an automated agent can be created, it is not easy to test the performance of the agent due to the complexity in supply chain caused by the interaction among the elements in it.

To simulate a supply chain model, TAC-SCM (Trading Agent Competition-Supply Chain Management) was suggested by a team of researchers from the e-Supply Chain Management Lab at Carnegie Mellon University, the University of Minnesota, and the Swedish Institute of Computer Science (SICS). The purpose of this competition is to capture many of the challenges involved in supporting dynamic supply chain practices, while keeping the rules of the game sufficiently simple [8]. This game is designed to simulate a B2B supply chain model of computer manufacturing company. However, the company in the simulation does not manufacture any computer parts. Instead, it purchases all of the parts from suppliers, assembles them, and sells completed products to customers. Hence, the supply chain model of this simulation is composed of suppliers, customers, and agents, and a game is regulated and operated by many detailed rules, which control the interactions of three main elements.

Considering this situation, we concluded that TAC-SCM is an appropriate framework to simulate and compare the performance of our software agent provided. Hence, we adopted this game as a basic framework for our simulation, but modified some rules of simulation to accommodate a more realistic business environment, to test the performance of our supply chain management agent.

2 Related Works

Since the beginning of the TAC SCM, numerous agents have joined the competition with various strategies. Among such agents, several agents such as DeepMaize, Tac-Tex, and MinneTAC, consistently showed better performance than other agents. However, due to the characteristics of the competition, many agents have used similar techniques for the optimization of the performance of each module and for the coordination of decision making among the modules in an agent.

For the component inventory management, many agents have used the concept of the inventory threshold. The advantage of this concept is that the agent may prevent the shortage of a component caused by a sudden increase of demand while it also has the drawback that it may cause higher procurement cost and higher holding cost. In spite of the drawback of the inventory threshold, most of the agents adopted this concept in their implementation. Agent A [17], MinneTAC [9], Metacor [6], DeepMaize [13], TacTex-06 [15], TacTex-03 [14], and RedAgent [10] employed this concept, although their specific implementation details are different.

For the production scheduling, a greedy algorithm is mainly used by Agent A [17], CiMeux [4], DeepMaze [13], TacTex-06 [15], TacTex-03 [14], and CrocodileAgent [16]. Botticelli [2, 3] used stochastic programming. For the reserve price in a procurement RFQ, various techniques are used. Agent A [17] and CrocodileAgent [16] setup their own formula for the calculation of reserve price. MinneTAC [9] and CiMeux [4] applied a nearest-neighborhood method and Metacor [6] used the most beneficial price observed by probes. DeepMaize [11] and TacTex-03 [14] track the previous three days' price and use price-capacity formula shown in [11] and TacTex [14].

In case of bidding price and the decision for whether to send the bid, a group of agents selected the family of linear programming: Agent A [17] used dynamic programing; MinneTAC [9] applied linear programming; Botticelli [2, 3] used stochastic programming; Foreseer [5] used a mixed linear programming; and CiMeux [4] used continuous Knapsack problem. The k-nearest neighbor algorithm is also adopted in CiMeux and Mertacor [6]. Search algorithm is employed in the following agents: binary search in CiMeux [4]; gradient descent search and a greedy algorithm in DeepMaize [13], and search of max increase of profit in TacTex [15], and a greedy algorithm for deciding to send bids and a linear regression for bidding price CrocodileAgent [16].

[9] provided an excellent insight for the classification of research agenda for an intelligent agent for supply chain management through an informal survey of the research teams. They suggested seven research issues as the result of the survey. We expect that these criteria would be used as a guide line for the automated SCM agent. Table 1 is the summary of the result.

Table 1. Research agenda and agents

Research agenda	Team
Constraint optimization	Botticelli, Cmieux, Foreseer, MinneTAC, DeepMaize
Machine learning	Cmieux, DeepMaize, MinneTAC, TacTex
Dynamic supply-chain	Cmieux, Foreseer, Mertacor, MinneTAC
Scalability	CrocodileAgent - IKB (Inforamtion Layer, Knowledge Layer, Behavioral Layer)
	MinneTAC - Blackboard Architecture
Empirical game theory	DeepMaize
Decision coordination	Cmieux, DeepMaize, Mertacor
Dealing with uncertainty	Botticelli, Foreseer, MinneTAC

In addition to TAC SCM literature, other studies on supply chain management agents were conducted. [19] provided the meta-learning mechanism algorithm for intelligent agents' learning parameters. In their study, the agent uses the Softmax method [20] to select an action (for e.g., choosing the bidding price) for their action value estimation function using the Eq. (1):

$$\Pr(a) = \frac{e^{Q(a)/\tau}}{\sum_{i=1}^{n} e^{Q(i)/\tau}} \qquad where \ \tau > 0 \qquad (1)$$

where a is the action that can be taken in action estimation function, $Q(a)$ is the action-value estimation function (i.e., the estimated value of taking action a), and τ is the temperature used for Softmax method. Also they employed ε-greedy algorithm to learn the temperature τ using the Eq. (2):

$$\Pr(a) = \frac{1-\varepsilon}{m} \quad if \ a = \underset{i}{argmax}(Q(i))$$
$$= \frac{\varepsilon}{k-1} \quad else \qquad (2)$$

[18] suggested the concept of Sliding Window Protocol for trading agent's strategy for bidding price. Instead of using fixed price band for bidding price selection, they provided flexing price band by using Sliding Window Protocol. The Sliding Window Protocol not only provides flexible price band, but also provides a mechanism for dynamic density control for the range of price band.

3 The Model

In this study, we designed a supply chain management simulation where multiple intelligent agents compete each other to make higher profit. Although this simulation is motivated by TAC SCM defined in [8], we redefined many parts to provide more realistic business environment. The simulation environment provides basic elements of simulation such as suppliers, customers, factory and storage of our agent, and other competing agents, and operates overall simulation events such as sending sales RFQs, sending procurement offers, moving next day, and so on.

3.1 Procurement and Suppliers

The simulation environment provides roles for suppliers. Among suppliers' tasks are those when suppliers need to issue procurement offers. They include making a decision for offer price, offer amount, and delivery date for each procurement offer. Procurement process begins when agents send procurements RFQs to suppliers. At this time they have to include component type, reserve price, quantity, and due date in their RFQs. Reserve price is a maximum price that agents may accept. Once procurement RFQs are received, each supplier groups the RFQs by the due date and sorts the RFQs in a group by the descending order of the reputations of agents.

Reputation is a mechanism used by suppliers to provide priority to more "loyal" agents. Each supplier then sets up offer price to the first offer for the agent having the highest reputation by using pre-defined supply price distribution. For agents with lower reputations, suppliers increase the offer price to give the advantage to the agents having higher reputation. After setting up the offer prices, each supplier sets up the quantity. Each supplier considers if it can supply entire quantity for each RFQ. If the supplier has enough capacity, a supplier will provide an offer having the same offer quantity to that in corresponding procurement RFQ. However, if the supplier doesn't have enough available capacity for RFQs, the supplier creates partial offers of which quantities are distributed among the group of RFQs having same due date.

As the final step of creation of procurement RFQs, suppliers create new offers that can complement the amount not filled with partial offers by finding the earliest possible day when each supplier can fill the deficit amount. For the partial offer, suppliers apply same offer price decided in the price setup step to continue to provide advantage to more loyal agents. After processing three steps, suppliers send all the procurement offers to the agents.

3.2 Sales and Customers

In this simulation, we designed customers so that they use pre-defined number of daily RFQs and daily demand amount for each PC type to manage the randomness and repeatability when creating sale RFQs. Customers make two types of decisions: sending customer RFQs and accepting sales offers.

For customer RFQs, customers first find the number of RFQs generated from a Poisson distribution. Customers then select due dates randomly between 3 and 12 day after the current day for each RFQ. Finally, they decide the RFQ quantity within the boundary of pre-defined demand distribution and send RFQs to agents. The agents then have to decide whether they will send sales offer to bid on the RFQ or not. After receiving the sale offers from agents, customers begin to consider sales offers to give final orders to agents. The first criteria is reserve price. If the bid price in a sales offer does not exceed the reserve price in the corresponding sales RFQ, the offer will be considered in the next step. Otherwise it will not be considered for sales orders. Once the reserve price constraint is met, customers select a sales offer that has the lowest bidding price for a sales order. If multiple agents bid the same lowest price, customer chooses one of the agents randomly and provides a sales order to that agent.

3.3 Competitors

In our simulation agents compete against each other to acquire suppliers' components and to win customers' sales demand. For evaluating the performance of our agent, we implemented two competitors as part of the simulation environment. As the first competitor, Agent-D is modeled by adopting the main strategies and algorithms used in DeepMaize in TAC SCM. Among the publications regarding DeepMaize, [11–13] are mainly referenced. For second competitor, we modeled Agent-T, referencing [14, 15] to employ the strategies and algorithms used for TacTex in TAC SCM [8].

3.4   The Modification from TAC SCM

Although this study is motivated by TAC SCM, we modified the simulation rules to implement more realistic business environment because there is no start, end, and data from previous simulation in real world business.

First, starting and ending conditions of the simulation are not considered by the agents. According to TAC SCM literature, agents can consider starting and ending conditions to enhance their performance. For example, TacTex-06 [15] decreases the component inventory threshold linearly to prevent the excessive inventory at the end of the simulation. However, in this study, no agent is allowed to use any algorithm for processing the initial status of the simulation or end of game situation. Although there are several rules for the end of the simulation in TAC SCM, in this study we just calculate the total benefit (or profit) of each agent as of the final simulation day to find "current" status of agents.

Second, most of the agents in TAC SCM make prediction data from previous rounds of competition and use it as a prediction tool for the current simulation. However, we don't provide any prediction data to the agents. Instead, we share the customer demand distributions and the suppliers' supply distributions with competing agents to allow them to have the highest prediction accuracy. However, our agent SCMaster, does not use any prediction data. These setup features provide the facilities for simulations in various environments because researchers may control the distribution of demand and supply by simply adjusting the supply and demand information defined in a file.

3.5   SCMaster

In this study we modeled an intelligent agent and named it SCMaster. Basically SCMaster adopted ideas used in previous agents. However, we created our own modular structure and applied various Reinforcement Learning techniques to improve the performance of sales offer and component inventory management. To deal with complex tasks in supply chain management, a modular approach is appropriate to balance the coupling of various modules while maintaining the independency among modules.

In our simulation, there are six different functions to be performed by agents every day, which are: (1) Send Procurement RFQs, (2) Receive Procurement Offers, (3) Send Sales Orders. (4) Receive Sales RFQs, (5) Send Sale Offers, and (6) Receive Orders. (1), (2), and (3) are tasks related to procurement and (4), (5), and (6) are related to sales. According to these categorization, we created three modules: Procurement Manager, Sales Manager, and Dashboard Manager. The role of Procurement Manager is to carry out all the tasks regarding the procurement of components from suppliers. Its sub tasks are: creating procurement RFQs, receiving offers from suppliers, making a decision whether to accept the procurement offers and sending orders to suppliers. A Sales Manager is responsible for all the tasks regarding customer demand. Its sub tasks are: receiving sales RFQs from customers, making a decision whether to bid on each sales RFQ at certain price and sending the bids to customers, accepting orders from customers and scheduling production and shipment.

A Dashboard Manager controls central data repository directly such as retrieving order status, procurement status, and inventory information and provides interface for data repository. Also, the Dashboard Manager provides actual implementation of business logic. Hence, the Supply Manager and the Procurement Manager may have abstracted view for data repository and business logic, and, at the same time, they can remove the limitation of data access range.

To improve the bidding performance we adopted Sliding Window Protocol suggested in [18] and adopted the concept of partial offer from [15]. Since the Sliding Window Protocol provides a mechanism for dynamic price band that can maintain the finer granularity for more popular price bands in agent's search space by consolidating the weaker price with coarser intervals, it provided improved predictions of winning probability for bidding prices. Hence, the agent is able to calculate more accurate partial offer amounts. In addition, we created a concept of profit threshold for sending a sales offer to focused on higher profit products. SCMaster filters sales offers having higher expected profit and only sends higher-profit offers. By combining the profit threshold and Sliding Window Protocol, the agent can improve the performance of sales offer because it can focus on the products having higher profit with higher winning probability.

We also improved SCMaster's procurement performance by creating dynamic inventory threshold. We adopted the concept of fixed threshold from previous agents, but made it dynamic by creating new equation and applying Reinforcement Learning. First, we created equation for dynamic inventory threshold in Eq. (3)

$$B_t^c = B_{t-1}^c + \alpha[(B_t^c - I_t^c) - (B_{t-1}^c - I_{t-1}^c)] \tag{3}$$

where B_t^c is the inventory threshold of component c on day t, I_t^c is the inventory level of component i on day t, and α is a learning rate. Then, we applied Softmax and ε-Greedy Reinforcement Learning techniques to make the learning rate α dynamic. By doing so, we can make inventory threshold shift dynamically. The advantage of adopting RL technique is the learning rate will explore when the performance of the Eq. (3) is worse and it will exploit the benefit when the performance of Eq. (3) is better.

4 Experiment and Analysis

4.1 Software for Experiment

For this study, a custom software was created to simulate the TCM SCM scenario. The software is implemented using Java 1.7, and a relational data base is used for the data repository of the agent. The software for this study has ability to run multiple instances of three types of agents, SCMaster, Agent-D, and Agent-T, as configured in starting parameters of a simulation. The agents in a simulation compete against each other to acquire customer sales and suppliers' components. For the competition among agents, this software provides a simulation environment including suppliers, customers, factory, and storage. The sever module also is responsible for operating the simulation flow such as moving the simulation to next day, sending signal for communication to all agents, creating daily and 20 day operation reports, and operating factory and delivery.

4.2   Experiment and Result

To test the performance of SCMaster, the agent competes with other types of agents, Agent-D and Agent-T, respectively in different simulations. For the statistical analysis, each simulation was executed 10 times. Wilcoxon's Matched-Pair Signed Ranks Test was used for the comparison of two agents' performance in all simulations. Tests for the matched-pair were used because different types of agents are executed and competed against each other within the same simulation. Also, to find the effect of different demand distributions, four different distributions were applied in the simulation 1 and the simulation 2. The experiment setting is depicted in Fig. 1.

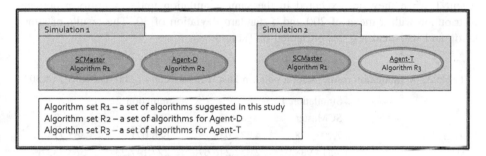

Fig. 1. Competitive heterogeneous agents

First, simulation 1 and 2 were conducted under a uniform demand distribution with a mean of 200 and a standard deviation of 10. After the 10 runs of simulation 1, the average profit of SCMaster was 7.76 M and its standard deviation was 0.32 M; the average profit of Agent-D was 4.44 M and the standard deviation was 2.00 M. The result was tested with Wilconxon's singed test, and the result showed the level of significance of .005. According to the result, it is concluded that SCMaster performed better than Agent-D when playing together in a same simulation under a uniform demand distribution with a mean of 200 and a standard deviation of 10. Under the same demand distribution this study conducted the simulation 2 and statistical test, and the test showed the level of significance at .005. Hence, one may concluded that SCMaster has better performance when executed with Agent-T in the simulation 2 under the uniform demand distribution with a mean of 200 and a standard deviation of 10. The results of the simulations and statistical tests are summarized in Table 2.

Table 2. A simulation result under a uniform Dist. with a mean of 200 and an SD of 10

	Simulation 1		Simulation 2	
	SCMaster	Agent-D	SCMaster	Agent-T
Avg.	7,760,469	4,442,391	7,645,224	4,398,388
SD	324,569.41	2,001,522.68	343,271.33	216,113.87
Wilcoxon's signed ranks test (2-tailed)	Sig. = .005		Sig. = .005	

For a uniform distribution with a mean of 200 and a standard deviation of 30, this study conducted the same set of simulations. The result of simulation 1 showed that the average profit of SCMaster was 7.90 M and its standard deviation was 0.31 M; the average profit of Agent-D was 4.27 M and the standard deviation was 2.33 M. Since the level of significance of Wilconxon's singed test was .005, it is concluded that SCMaster performed better than Agent-D when playing together in a same simulation under the uniform demand distribution with a mean of 200 and a standard deviation of 30. The result of simulation 2 showed that the average profit of SCMaster was 7.69 M and its standard deviation was 0.23 M; the average profit of Agent-T was 4.37 M and the standard deviation was 0.36 M. According to the result of Wilcoxon's signed rank test (Sig. = .005), one may conclude that SCMaster has better performance than Agent-T when they are executed in the same simulation under a uniform demand distribution with a mean of 200 and a standard deviation of 30. The results of simulation and statistical test are presented in Table 3.

Table 3. A simulation result under a uniform Dist. with a mean of 200 and an SD of 30

	Simulation 1		Simulation 2	
	SCMaster	Agent-D	SCMaster	Agent-T
Avg.	7,903,478	4,271,035	7,688,071	4,365,773
SD	314,988.03	2,332,246.67	227,147.23	363,332.69
Wilcoxon's signed ranks test (2-tailed)	Sig. = .005		Sig. = .005	

For two other distributions, a uniform distribution with a mean of 200 and a standard deviation of 50 and a uniform distribution with a mean of 200 and a standard deviation of 100, this study conducted same simulations and statistical tests. The results of all statistical tests are the same as those in previous tests. In simulations 1 and 2, the SCMaster showed better performance than Agent-D and Agent-T, respectively. The results are presented in Tables 4 and 5 and Figs. 2 and 3. As a result of the simulations 1 and 2, we can conclude that the SCMaster has better performance than other two agents in all demand distributions.

Table 4. A simulation result under a uniform Dist. with a mean of 200 and an SD of 50

	Simulation 1		Simulation 2	
	SCMaster	Agent-D	SCMaster	Agent-T
Avg.	7,508,036	2,804,434	7,494,287	4,435,694
SD	254,793.03	1,968,954.50	439,093.35	227,682.38
Wilcoxon's signed ranks test (2-tailed)	Sig. = .005		Sig. = .005	

Table 5. A simulation result under a uniform Dist. with a mean of 200 and an SD of 100

	Simulation 1		Simulation 2	
	SCMaster	Agent-D	SCMaster	Agent-T
Avg.	8,156,249	3,750,077	8,037,760	4,855,926
SD	402,246.26	2,525,082.47	505,412.31	228,041.59
Wilcoxon's signed ranks test (2-tailed)	Sig. = .005		Sig. = .005	

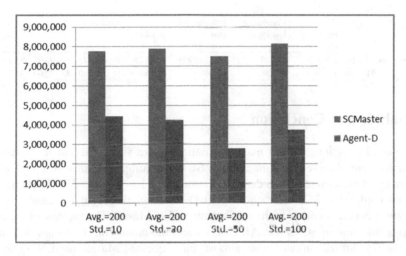

Fig. 2. Performance comparison of SCMaster in simulation 1

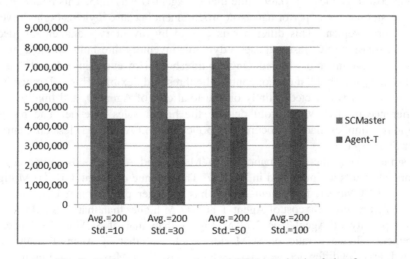

Fig. 3. Performance comparison of SCMaster in simulation 2

Table 6. Summary of performance of the two agents in simulation 1

	Revenue	Cost				Total cost	Profit	Factory utilization
		Product holding cost	Component holding cost	Procurement cost	Sales penalty			
SCMaster	98,626,474	142,230	396,037	88,130,125	1,973,074	90,641,465	7,985,009	97.0 %
Agent-D	104,476,150	116,402	352,096	97,977,200	1,841,372	100,287,071	4,189,079	97.3 %

Table 7. Summary of performance of three agents in simulation 2

	Revenue	Cost				Total cost	Profit	Factory utilization
		Product holding cost	Component holding cost	Procurement cost	Sales penalty			
SCMaster	98,302,664	136,337	346,294	87,423,875	1,546,267	89,452,773	8,849,891	95.9 %
Agent-T	100,945,820	117,280	409,588	91,246,725	4,256,666	96,030,260	4,915,560	97.2 %

5 Analysis and Conclusion

In the previous section, various market situations were simulated and the results of simulations were observed. In simulation 1, SCMater compete against Agent-D where their strategies are different. The details of simulation 1 is presented in Table 6.

When comparing SCMaster and Agent-D in simulation 1, it is found that the procurement cost was critical to the performance of SCMaster. The revenue of Agent-D was larger than that of SCMaster. All costs except the procurement cost are almost the same. Factory utilization was also almost the same: SCMaster used 97.0 % and Agent-D used 97.3 %. However, Agent-D paid about 97 M as procurement costs while SCMaster paid 88 M. In the case of SCMaster, the procurement efficiency (= procurement cost /revenue) was 0.89 while that of Agent-D was 0.93. This result showed that SCMaster has better performance on procurement because it yields higher revenue with less procurement. This difference is caused by inventory management mechanisms. In case of the SCMaster, it uses a dynamic inventory threshold, so the agent can prevent excessive inventory by reducing the threshold. However, since Agent-D uses a fixed inventory threshold, it cannot control the threshold flexibly. This difference in the procurement cost is reflected entirely on the total cost of Agent-D, and finally caused the difference of 3.79 M in profit despite the higher sales revenue. The result of simulations 1 implies that the procurement policy used in Agent-D is less efficient than that for SCMaster.

In simulation 2, the performance of SCMaster and Agent-T are compared. The summarized results are presented in Table 7. The revenue of Agent-T is 2.6 M higher than that of SCMaster, but Agent-T also has a higher procurement cost and sales penalty. The procurement cost of Agent-T is 3.82 M higher than that of SCMaster, and the sales penalty of Agent-T is 2.71 M higher than that of SCMaster. Hence, the procurement efficiency ratio of SCMaster, 0.89, and that of Agent-T, 0.90, were compared, and the difference was 0.01, which caused 1.1 M of profit difference.

This difference is caused by the inventory management mechanisms. As mentioned in the analysis of simulation 1, SCMaster uses a dynamic inventory threshold and Agent-T uses a fixed inventory threshold. Another weakness of Agent-T is the high penalty amount. Both agents use more than 95 % of the factory cycle on average (Agent-T = 1945 cycles/day = 97.2 %, SCMaster 1919.9 cycles/day = 95.9 %), but Agent-T won 7940 orders and SCMaster won 3447 orders. This implies that Agent-T has too many orders for the agent to process, and this excessive orders resulted in a high sales penalty. To the contrary, SCMaster has less excessive sales orders, and it created less of a sales penalty than that of Agent-T.

From this analysis, we may conclude that SCMaster's bidding strategy using a slide window protocol works effectively so that it increases the revenue while focusing on higher profit products and maintaining appropriate level of sales orders for less sales penalty. At the same time, the flexible inventory threshold also increases the procurement efficiency, effectively supplying the component for production.

References

1. Acquity Group. Uncovering the Shifting Landscape in B2B Commerce. 2014 State of B2B Procurement Study (2014). http://www.acquitygroup.com/news-and-ideas/thought-leadership/article/detail/2014-b2b-procurement-study
2. Benisch, M., Greenwald, A., Grypari, I., Lederman, R., Naroditskiy, V., Tschantz, M.: Botticelli: a supply chain management agent designed to optimize under uncertainty. ACM SIGecom Exch. **4**(3), 29–37 (2004)
3. Benisch, M., Greenwald, A., Naroditskiy, V., Tschantz, M. C.: A stochastic programming approach to scheduling in TAC SCM. In: 5th ACM Conference on Electronic Commerce, pp. 152–159. ACM, May 2004b
4. Benisch, M., Sardinha, A., Andrews, J., Ravichandran, R., Sadeh, N.: CMieux: adaptive strategies for competitive supply chain trading. Electron. Commerce Res. Appl. **8**(2), 78–90 (2009)
5. Burke, D. A., Brown, K. N., Hnich, B., Tarim, A.: Learning market prices for a real-time supply chain management trading agent. In: Workshop on Trading Agent Design and Analysis/Agent Mediated Electronic Commerce, AAMAS 2006 (2006)
6. Chatzidimitriou, K.C., Symeonidis, A.L., Kontogounis, I., Mitkas, P.A.: Agent Mertacor: a robust design for dealing with uncertainty and variation in SCM environments. Expert Syst. Appl. **35**(3), 591–603 (2008)
7. Choi, T.Y., Dooley, K.J., Rungtusanatham, M.: Supply networks and complex adaptive systems: control versus emergence. J. Oper. Manage. **19**(3), 351–366 (2001)
8. Collins, J., Arunachalam, R., Sadeh, N., Eriksson, J., Finne, N., Jansonl, S.: The supply chain management game for the 2007 trading agent competition (2006). http://tradingagents.eecs.umich.edu/
9. Collins, J., Ketter, W., Gini, M.: Flexible decision control in an autonomous trading agent. Electron. Commerce Res. Appl. **8**(2), 91–105 (2009)
10. Keller, P.W., Duguay, F.O., Precup, D.: Redagent-2003: an autonomous market-based supply-chain management agent. In: Third International Joint Conference on Autonomous Agents and Multiagent Systems, vol. 3, pp. 1182–1189. IEEE Computer Society, July 2004

11. Kiekintveld, C., Wellman, M.P., Singh, S., Soni, V.: Value-driven procurement in the TAC supply chain game. ACM SIGecom Exch. **4**(3), 9–18 (2004)
12. Kiekintveld, C., Wellman, M.P., Singh, S.P., Estelle, J., Vorobeychik, Y., Soni, V., Rudary, M.R.: Distributed feedback control for decision making on supply chains. In: ICAPS, pp. 384–392, June 2004b
13. Kiekintveld, C., Miller, J., Jordan, P. R., Wellman, M. P.: Controlling a supply chain agent using value-based decomposition. In: 7th ACM Conference on Electronic Commerce, pp. 208–217. ACM, June 2006
14. Pardoe, D., Stone, P.: TacTex-03: a supply chain management agent. ACM SIGecom Exch. **4**(3), 19–28 (2004)
15. Pardoe, D., Stone, P.: An autonomous agent for supply chain management. In: Handbooks in Information Systems Series: Business Computing, vol. 3, pp. 141–172 (2009)
16. Podobnik, V., Petric, A., Jezic, G.: The CrocodileAgent: research for efficient agent-based cross-enterprise processes. In: Meersman, R., Tari, Z., Herrero, P. (eds.) OTM 2006 Workshops. LNCS, vol. 4277, pp. 752–762. Springer, Heidelberg (2006)
17. Sibdari, S., Zhang, X.S., Singh, S.: A dynamic programming approach for agent's bidding strategy in TAC-SCM game. Int. J. Oper. Res. **14**(2), 121–134 (2012)
18. Sikora, R.T., Sachdev, V.: Learning bidding strategies with autonomous agents in environments with unstable equilibrium. Decis. Support Syst. **46**(1), 101–114 (2008)
19. Sikora, R.: Meta-learning optimal parameter values in non-stationary environments. Knowl. Based Syst. **21**(8), 800–806 (2008)
20. Sutton, R.S., Barto, A.G.: Reinforcement Learning: An Introduction. MIT Press, Cambridge (1998)

Investigating IT Standardization Process Through the Lens of Theory of Communicative Action

Karthikeyan Umapathy[1]([⊠]), Sandeep Purao[2], and John W. Bagby[3]

[1] University of North Florida, 1 UNF Drive, Jacksonville, FL 32224, USA
k.umapathy@unf.edu
[2] Bentley University, 175 Forest Street, Waltham, MA 02452, USA
spurao@bentley.edu
[3] Penn State University, IST Building, University Park, PA 16802, USA
jbagby@ist.psu.edu

Abstract. Developing standards is a social practice wherein experts engage in discussions to evaluate design solutions. In this paper, we analyze processes followed to develop SOAP standard from the theory of communicative action perspective, which argues that individuals engaged in social discourse would exhibit five possible actions: instrumental, strategic, normatively regulated, dramaturgical, and communicative. Our findings reveal that participants in standardization processes engage in communicative action most frequently with aim of reaching mutual understanding and consensus, engage in strategic action when influencing others towards their intended goals, engage in instrumental action when taking responsibility for solving technical issues, engage in dramaturgical action when expressing their opinions, and engage in normatively regulated action when performing roles they assumed. Our analysis indicates that 60 % of activities performed are consensus oriented whereas the rest are success oriented. This paper provides empirical evidence for Habermasian view of social actions occurring in the standardization process setting.

Keywords: Standards · Standardization processes · Habermas · Communicative action

1 Introduction

Anticipatory standards are created before the technology is introduced in the market. Anticipatory standards are developed to ensure continued evolution of Information Technology (IT) landscape and are compatible with existing technologies. Thus, anticipatory standards provide technical specifications for developing new products and services. Examples of anticipatory standards include HTML (HyperText Markup Language) and SOAP (Simple Object Access Protocol). Developing anticipatory standards are difficult as it involves substantial design components that are reviewed and approved by industry members. Anticipatory IT standards face the threat of being premature and full of unnecessary details [1], when inappropriate compromises are made during standards development process. An inapt anticipatory standard can have

© Springer International Publishing Switzerland 2016
V. Sugumaran et al. (Eds.): WEB 2015, LNBIP 258, pp. 40–53, 2016.
DOI: 10.1007/978-3-319-45408-5_4

detrimental effect on the industry as it can be considered irrelevant, fail to aid in progressing the industry, or force the industry into a dead-end with poorly conceived design specifications [1].

Most IT standards, particularly Web standards, are developed by consortium-based Standard Development Organizations (SDOs) like W3C (World Wide Web Consortium) rather than traditional SDOs like ISO (International Organization for Standardization). Consortium-based SDOs have an advantage over traditional SDOs, in that, participating members may exhibit greater tendency to be like minded, willingness to cooperate, and wish for standard to be created. Consortium-based SDOs provide a forum for interested parties to share their innovations and develop mutually agreed standards.

Consortium-based SDOs have placed safeguards to ensure their process is open but not abused. For an example, W3C has issued a process document [2] that provides guidelines to regulate processes followed to develop standards. Apart from openness, success of consortium processes rely on voluntary contributions from its members, and participants reaching consensus when making critical decisions relevant to the standard. Due to the openness, consensus orientation, and volunteer participation, many researchers have argued that standardization processes are quite similar to Habermasian view of rational discourse (i.e., open-ended discussion geared towards reaching consensus) described in the theory of communicative action [3–6]. However, none have conducted empirical investigation on an actual standardization process to provide evidence of social actions described by Habermas occurring within the process. Thus, the objective of this paper is to investigate IT standardization process from the theory of communicative action perspective and find evidence of social actions within an actual standardization process.

To aid our investigation, we conduct a qualitative study on the process followed to develop SOAP standard. We performed content analysis on transcripts of SOAP committee meetings to extract commonly performed activities. These activities were further analyzed using Habermas's theory of communicate action [7] to identify activities that show evidence of occurrence of social actions within the standardization process. We anticipate that our analysis from the theory of communicative action perspective would provide a richer insight into inner workings of the standardization process that is largely unknown and under-researched.

2 Background

SDOs play a vital role in developing IT standards by bringing together experts from multiple industry sectors to collaboratively work towards a common solution for complex problems relevant to the industry. Traditional SDOs develop de jure standards based on a regulatory process which can be bureaucratic and time consuming [8]. Thus, most IT standards are developed by consortium-based SDOs as voluntary standards which are similar to de facto standards but developed through coordination among and voluntary contribution from consortium members [9]. IT standards contain significant technical component which are designed collaboratively by vested parties who are competitors in the market place. Even though IT standards developed by consortium-based SDOs play central role in advancing and shaping future direction of the field, we lack adequate understanding on how IT standards are designed and developed [10]. There are other prior works focusing on vertical industry specific e-business standards [9] and comparative study on business process centric and technology centric e-business standards [10]. However, in this paper, we focus on anticipatory IT standards, i.e., standards developed prior to development of the technology. Our research objective is to investigate ability of theory of communicative action using empirical evidence through conducting grounded analysis on standardization processes followed at W3C.

2.1 Theory of Communicative Action

According to Habermas [7], a society maintains itself through socially coordinated activities of its members and that this coordination is established through communication aimed at reaching consensus (p. 397). Habermas argues that a democratic system of exchange among members is crucial for supporting communication that are aimed at reaching consensus [11]. Habermas [7] suggests that members who may have conflicting interests and working towards a common goal can reach consensus through an argumentative process where statements of each member are verified and validated (p. 86). Following Habermas arguments, we argue that standardization processes should provide an open forum for all participants to engage in a social discourse and have equal opportunity to initiate and sustain communication; rather than a process that is riddled with authority, tradition, power, or prejudices.

Habermas, in the theory of communication action [7], provides a theoretical framework that can be used for analyzing processes that are consensus oriented and involve open participation and information sharing through dialogue as opposed to exercising power of authority [12]. Thus, consensus oriented processes such as IT standardization processes can be analyzed using assumptions, conditions, and principles provided in the theory of communicative action to gain a better understanding of inner working of these processes [6].

There is considerable literature containing detailed overview of theory of communicative action [4, 12–15]. Thus, in this section we provide a brief overview on Habermas's classification of social actions performed by individuals in a society. Habermas developed social action categories based on human predispositions of striving to succeed based on available resources and power, and reaching mutual understanding through coordination of actions with partners [16]. He distinguishes five kinds of social actions: instrumental, strategic, normatively regulated, dramaturgical, and communicative. Instrumental and strategic actions are geared towards achieving success, while normatively regulated, dramaturgical, and communicative actions are geared towards achieving mutual understanding [17]. Table 1 provides discourse objective and action orientation for the five social actions.

Instrumental Action. The main concept of instrumental action is about making decisions on alternative courses of action to achieve the goal based on an interpretation of the given context [4]. Thus, instrumental action is performed by an actor to achieve a goal or success, which is measured by effectiveness in achieving the desired objective.

Table 1. Five types of social actions.

Social actions	Discourse objective	Action orientation
Communicative action	Coordinating of the plans of individual participants to achieve common goals	Consensus-orientation
Strategic action	Influencing other participants	Success-orientation
Instrumental action	Representation of states of affairs	Success-orientation
Dramaturgical action	Self-representation	Consensus-orientation
Normatively regulated action	Establishing and following norms	Consensus-orientation

Individuals performing instrumental actions would manipulate physical objects, abstract entities, states of affairs, or other humans to achieve their goals [15]. These manipulations are performed through application of technical rules derived from experiential knowledge [4, 16]. Furthermore, application of technical rules involves selection of an action from a set of alternatives that maximize their chances to achieve desired objectives [15].

Strategic Action. Habermas refers strategic and instrumental actions as teleological actions, which involve following certain set of decision rules to maximize actor's interest or goal. Strategic action differs from instrumental action in that one actor perceives other actors not as an object but as a rational opponent and player in the game [15]. Individuals performing strategic actions would undertake a purposeful rational action while assessing and anticipating reactions of other participants [1, 12, 17]. Depending upon the social context and whether other actors' goals coincide with or oppose to their, individuals would make a rational choice among alternatives to maximize chances of achieving a desired objective [6].

Normatively Regulated Action. Habermas refers actions performed in a social group in accordance to established common values or norms as normatively regulated action [12]. Unlike instrumental and strategic actions, normatively regulated actions are performed automatically in rote fashion as opposed to manipulative manner [12, 15]. Furthermore, normatively regulated actions are performed by actors in their social roles during interaction with other members of the group [12]. Normatively regulated actions can be assessed based on their normative rightness and legitimacy [15].

Dramaturgical Action. An individual's act of self-expressing to present certain image or impression of themselves is referred as a dramaturgical action [12]. Dramaturgical actions are performed in public forum while interacting with a social group. Dramaturgical actions include presentations of an actor's identity, interests, strengths, and weakness. Expressions of the actor are evaluated by other members on the intended meaning of the presented position [18].

Communicative Action. Engagement of two or more actors to reach mutual understanding and coordinated plan of actions to maintain mutual agreement is known as communicative action [12]. Actors involved in communicative action reach mutual understanding and consensus by conducting argumentative discourse about norms, common values, states of affairs, events, decisions taken and so forth [15, 16]. Communicative action can be considered successful if consensus is reached through cooperative manner and through coordinated plan of actions by involved actors [15]. If there exists misunderstanding among actors and difference in opinions, goals, or plans of actions, then communicative action breaks down [16]. When breakdown occurs, actors engage in civilized argumentative process known as rational discourses to reach consensus and find a common ground [16, 17].

Openness and consensus orientation are the cornerstone of standardization processes. Thus, the theory of communicative action is an appropriate framework for investigating IT standardization processes. In the next section, we provide details of categorical standardization activities identified through application of content analysis on a SOAP standardization process. In the following section, we analyze these categorical activities against the five social actions provided by the theory of communicative action.

3 Case Study: SOAP Standardization Process

We employed a case study approach [19] to examine anticipatory IT standardization processes. For the purpose of this examination, we selected SOAP (Simple Object Access Protocol) 1.2 [20] standard developed by W3C. SOAP defines an XML-based framework for exchanging information among web services [21]. SOAP is an anticipatory standard as it contains substantial design components and development of SOAP paved way for web service technologies.

We used content analysis technique [22] to inspect SOAP standardization processes. Content analysis technique can be applied on large amounts of texts to make categorical inferences on a phenomena in a methodical and repeatable manner [23]. Content analysis technique was applied on meeting transcripts (face-to-face and telephone meetings) which are publicly available documents at the W3C XML working group committee site [20]. There are a total of 120 meeting transcripts that archive process followed to develop SOAP 1.2 standard. We examined SOAP meeting transcripts without any preconceived notion. Thus, along with content analysis, we also applied grounded theory approach to identify trends and patterns of activities that may have occurred during the development of SOAP standard.

A research team performed content analysis on the meeting transcripts. Each member in the team performed coding as well as questioning, examining, and corroborating interpretations made by other members. Coding in the context of content analysis involves identifying a text fragment and assigning appropriate categorical inferences. Categorical inference expresses contextual meaning attached to the identified text fragment in relevance to the phenomenon under study. In this research context, a categorical inference relates to a significant event or an activity performed by participants of SOAP standardization. Atlas.Ti, a content analysis software [24], was used for managing meeting transcripts and annotating categorical inferences obtained from specific text fragments in the documents.

In regards to validity of categorical interference made using content analysis, establishing high inter-rater reliability is critical. Inter-rater reliability represents the extent to which different coders make same categorical inference when each coding the same text fragment [25]. As content analysis involves different research team members reading and inferring the meaning of the text, inter-rater reliability is a test assessment of objectivity. Inter-rater reliability measures provided by Neuendorf [26] were used to calculate inter-coder reliability of the research team. To establish inter-rater reliability, the analysis was initially performed independently by the coders on the same set of documents. The resulting categories from each coder were compared to ensure that they had established a shared understanding of unit of analysis (text fragments to be identified), and category interfered (assigned meaning). Inter-rater reliability of 81 % was achieved after three iterations (for each iteration, three randomly selected documents were used for analysis). For exploratory studies similar to this research, inter-rater reliability above 70 % is considered to be adequate [27].

To ensure consistent interpretation of text fragments, coding scheme consisting of syntactic (unit of analysis) and semantic (categorical inferences) rules was developed. After establishing the coding scheme, coders independently performed content analysis on rest of the documents. Although the coding scheme emerged through inter-rater reliability process, new codes were allowed to emerge throughout the analysis. When a new code emerged, coders regrouped to ensure they have common understanding and interpretation, subsequently coding scheme was updated to include new code.

Empirical analysis all of 120 meeting transcripts that archived SOAP processes revealed 95 purposeful activities performed by participants. Of 95, 39 activities were main categorical activities while rest were sub-categories that are semantical variant to the main categories. An example of sub-category variants would be, action items as a main category and action item performed by large, small, chair, and W3C would be sub-categories. In the next section, we analyze 95 activities inferred from content analysis and map them to five social actions based on the theory of communicative action framework.

4 Analysis Using Theory of Communicative Action

To aid our analysis, activities emerged from content analysis of SOAP standard were mapped to five social actions described in the theory of communicative action. Mapping was performed based on the meaning assigned to activities inferred during content analysis and description of the social actions provided in the theory of communicative action. In order to achieve reliable and objective mapping, each researcher independently mapped each of the 95 inferred activities to one of the five social actions. For activities with contradicting mappings, researchers engaged in discussion until mutual agreement was reached. Application of the social actions to the inferred activities revealed that most of the activities can be mapped to five social actions explicated by Habermas, except for two, which are text fragments in the meeting minutes being censored and usage of both assistance to remind participants about important issues of the meeting. Both bot assistance and censored activities occurred only once during the SOAP standardization process.

Application of the Habermas's social actions indicated that out of 95 inferred activities—10 were instrumental action activities, 26 were strategic action activities, 13 were normatively regulated action activities, 18 were dramaturgical action activities, 26 were communicative action activities, and 2 activities did not fit with any of the five social actions. Overall 98 % of inferred activities can be mapped to social actions. It indicates that the theory of communicative action is relevant to the standardization process followed by W3C for the developing SOAP standard.

Further analysis was performed by totaling text fragments annotated (4760) with inferred activities against to five social actions. Table 2 provides counts and percentages of social actions and annotated activities mapping. The analysis indicates that communicative action is at the core of the standardization process, which is not a surprising finding because reaching consensus is central to the process. Analysis also shows that strategic action is the second most frequent social action, followed by instrumental and then by dramaturgical action activities. Normatively regulated action is the least frequent, which indicates that most participants followed established norms and spent least amount of time on establishing new norms.

Table 2. Social action count for all activities annotated.

Social Actions	No. of inferred activities	No. of text fragments	% of fragments
Communicative action	26	1507	31 %
Strategic action	26	1033	22 %
Instrumental action	10	858	18 %
Dramaturgical action	18	705	15 %
Normatively regulated action	13	656	14 %

To gain a better understanding on what kinds of inferred activities are associated with each social actions, we tallied counts at the sub-categories level for each social action. Following sub-sections provide discussions on each social action.

4.1 Instrumental Actions

The goal-oriented instrumental action is performed by participants when their objective is to advance their own personal interests. Participants seeking to obtain desired results would utilize their technical knowledge accrued through experience. Table 3 provides top ten most frequent instrumental action activities. From the table, it can be noted that action items (to solve technical/design problems) identified by specific participants and participants volunteering to solve a technical issue are top two main categorical activities. Identifying action items by specific participants shows that participants are identifying technical issues to be resolved to ensure the standards meet their interests and goals. Participants are also volunteering to resolve technical issues based on their skills gained from their work experience or availability of resources from their organizations. Thus, participants engaging in instrumental actions achieve their goals by identifying technical issues and by choosing to work on technical issues that would allow them to achieve their intended goals for participating in the standards development process.

Table 3. Ten most frequent instrumental action activities.

Activity	Frequency count
Identifying action items to be performed: Large Organization	339
Identifying action items to be performed: Small Organization	131
Identifying action items to be performed: Editors	109
Identifying action items to be performed: Chair	95
Identifying action items to be performed	67
Identifying action items to be performed: Responsibility W3C	60
Volunteering to perform an activity: Large Organization	44
Volunteering to perform an activity: Small Organization	7
No Volunteers	3
Volunteering to perform an activity: W3C	3

4.2   Strategic Actions

The purposive strategic action is carried out with contemplation of other participant's reactions. Participants engaging in strategic action aim at influencing other participant's actions. Table 4 provides top ten most frequent strategic action activities. From the table, it can be noted that providing design (substantial design contribution), suggesting design (minor design contribution), and raising issues with design solutions are top three main categorical activities. Providing and suggesting design solutions can be considered as strategic actions as participants are presenting alternative solutions for other participants to consider. While proposing design options, participants are assessing needs and goals of the standard, and presenting an option that will sway other participants towards their desired solution option. Similarly, raising issues in existing solution, raising overlap among standards, and raising overlap among other issues can be considered as presenting information on how existing solution can negatively affect the standard. Thus, through raising issues, participants are trying to sway other participants from a proposed solution. From the Table 4, it can be also noted that some participants can strategically choose to procrastinate from proposing solution for assigned action item, thus, buying more time to identify a solution that meets their goals and needs. Thus, participants engaging in strategic actions would present solution options or raise issues with existing solution options to influence other participants towards their intended goals within the standard.

Table 4. Ten most frequent strategic action activities.

Activity	Frequency count
Suggesting alternative design solutions: Large Organization	197
Raising issue with design solutions: Large Organization	176
Proposing design solutions: Large Organization	167
Suggesting alternative design solutions: Small Organization	68
Procrastination	67
Suggesting alternative design solutions: Chair	47
Standard Overlap Information	45
Raising issue with design solutions: Small Organization	37
Proposing design solutions: Small Organization	36
Issue Overlap information	34

4.3   Normatively Regulated Actions

Normatively regulated action is performed by participants when they act in accordance to the expected norms and values of the process. Participants engaging in normatively regulated actions would be either following established norms or establishing new norms for others to follow. Table 5 provides top ten most frequent normatively regulated action activities. From the table, it can be noted that most of the activities represent participants playing their excepted roles in the process. Roles such as chair and W3C staff are similar to assigned positions within the process, thus, they have prescribed norms to follow. However, other roles such as issue owners are voluntarily assumed by participants depending upon the context of ongoing discussions. As standardization is a social process, participants are expected to enact their roles within the process, follow associated norms, and perform actions that are in the best interests of the standard. Reporting progress action represents expectation from participants to report progress on the technical issues assigned to them. During each meeting, at the beginning participants are expected to report progress they made on the tasks assigned to them. Regulating process represents action performed (mostly by chair) to ensure meeting discussion is proceeding in the expected direction and completed in a timely fashion. Activities such as establishing and suggesting norms represent participants creating new norms to be followed by other participants. When established norms are not helpful in governing discussions within the meeting, participants engage in discussions to create new norms to handle the issue. Activities such as request for volunteer is performed by chair when participants do not volunteer to take responsibility for some open tasks. When participants do not volunteer on their own as expected, then the chair directly assigns action items to specific participant who may have the required technical skills. Thus, participants engaged in normatively regulated actions would be performing roles they assume within the process and expected actions associated with the assumed roles.

Table 5. Ten most frequent normatively regulated action activities.

Activity	Frequency count
Role: Chairperson of the committee performing his/her duties	240
Reporting Progress	151
Regulating Process	132
Request for Volunteer	51
Role: Owner of an issue	32
Suggesting Norm	18
Role: W3C staff	11
Establishing Norm	7
Role: Intermediator representative from another standard	7
Role: Proxy for a participant	4

4.4 Dramaturgical Actions

Dramaturgical action is performed by participants when they express their position on the issue under discussion. Participants engaged in dramaturgical action would be expressing their thoughts and interests with the standard. Table 6 provides top ten most frequent dramaturgical action activities. From the table, it can be noted that participants expressing their agreement with statements made by other participants is most frequent dramaturgical action activity. We also note that participants also express their confusion and disagreement of statements made by participants as well as show appreciation to other participants. Other frequent activities are expressing frustration and concerns with the process. Thus, participants engaged in dramaturgical actions would be expressing opinions on statements made by others and about the standardization process.

Table 6. Ten most frequent dramaturgical action activities.

Activity	Frequency count
Expressing agreement	539
Expressing concerns with the standard: Large Organization	52
Disagreement	51
Expressing reasons to participant in process: Large Organization	23
Expressing confusion with the discussions: Large Organization	17
Expressing concerns with the standard: Small Organization	7
Expressing appreciation	4
Expressing concerns with the standard: W3C	4
Expressing concerns with the standard: Chair	2
Expressing confusion with the discussions: Chair	2

4.5   Communicative Actions

Communicative action is performed by participants when they work towards reaching mutual understanding through coordination of their actions. Participants engaged in communicate action would attempt to reach consensus on the issue under discussion through a communicative process that is open for exchanging information, cooperation, and criticism. Table 7 provides top ten most frequent communicative action activities. From the table, it can be noted that participants most often engage in discussions on technical issues, W3C process, or political nature of standard development. Second most frequent activity is participant resolving a technical issue by reaching consensus on an appropriate solution for the issue. Most often to reach a resolution participants vote on various design options proposed for the issue. This voting process to resolve issues is reflected in the democratic process activity. Before voting on available design options, participants hold lengthy open-ended discussions wherein proposed solutions are critiqued and additional information is provided to add clarity to the solutions. This engagement in open discussions is reflected in the interaction among participants activity. When questions on issues cannot be answered or solutions cannot be developed individually by a participant, they agree to cooperatively develop a solution. Willingness and mutual agreement among participants to cooperatively accomplishing a task is reflected in cooperation among participants activity. Thus, participants engaged in communicative action would be aiming to reach mutual understanding and consensus through discussions and sharing information.

Table 7. Ten most frequent communicative action activities.

Activity	Frequency count
Discussions on technical issues	802
Issue Resolution	259
Discussions on W3C process	132
Democratic voting process to reach consensus	115
Interaction among Large Organizations	52
Cooperation for solving issue: between Large Organizations	36
Discussions on political issues associated with standard	25
Interaction between Large Organization and Small Organization	22
Cooperation for solving issue: Large and Small Organizations	16
Requesting for more clarity on existing design solutions	12

5   Discussions

Empirical analysis presented in this paper shows that standardization process followed at W3C exhibits all of the social actions described in the theory of communicative action. Habermas argues that an individual engaged in rational discussions would be either success oriented (focused on achieving individual goals) or consensus oriented (achieving mutual understanding) [7]. Success oriented individuals would perform instrumental and strategic actions, and consensus oriented individuals would perform normatively regulated, dramaturgical, and communicative actions. Analysis indicates that 60 % of activities performed by participants are consensus oriented and the rest are success oriented activities. Standardization process is a coopetition process as participating organizations are competing in the marketplace but cooperating to develop a standard as a common good for the industry. Our findings thus shed some light on how their actions split between cooperating (consensus oriented) and competing (success oriented).

Our analysis reveal that participants achieve consensus by conducting discussions on technical issues as well as about the process itself. Participants reach mutual understanding on the most viable design options by holding lengthy discussions. During such discussions, participants express their interests and concerns, as well as critique proposed solutions. When they cannot reach a unanimous decision on proposed options, they engage in a democratic process and select a design option by conducting a straw poll for each presented option. Chairperson of the committee plays an important role of regulating the process and ensuring participants are following established norms. Our findings reveal that participants engage in activities that involves contributing design options or assessing proposed solutions when they are success oriented. Participants engaging in success oriented activities rely on their technical capabilities and research investments of their organizations. Success oriented participants take on technical issues that need solutions, propose solutions for open issues, and assess the merits of proposed solutions.

To best of our knowledge, this is the first paper providing empirical evidence of occurrence of Habermasian view of social actions in the anticipatory standardization process setting. However more is left to explore in regards to IT standardization process and understanding it from Habermasian perspectives. Habermas argues that when there is misunderstanding, contradictory opinions, different viewpoints of the goals, or disagreements on best means to achieve a goal, then communicative action breaks down [7]. In situations when communication breaks down, participants would engage in a civilized argumentation process, known as rational discourse to find a common ground. There are five different types of argumentative discourses participants may engage in to overcome communication breakdown. As a part of future work, we intend to analyze inferred activities from our empirical analysis to observe when and how these argumentative discourses are manifested within the standardization process.

This study certainly has its limitations. In this study, we perform analysis only on one standard (i.e., SOAP) and on one SDO (i.e., W3C). Thus, findings from this study cannot be generalized for all anticipatory standards or SDOs. We need to analyze more standards from variety of SDOs. We need to study both successful and failed standards to gain full understanding of inner workings of these processes. Only then we will have necessary insights to improve the anticipatory standardization process efficiency and effectiveness.

References

1. Sherif, M.H.: A framework for standardization in telecommunications and information technology. IEEE Commun. Mag. **39**, 94–100 (2001)
2. W3C-Process: W3C. http://www.w3.org/2014/Process-20140801/
3. Schoechle, T.: Toward a theory of standards. In: IEEE Conference on Standardisation and Innovation in Information Technology (SIIT), Los Alamitos, CA, USA. IEEE (1999)
4. Lyytinen, K., Hirschheim, R.: Information systems as rational discourse: an application of Habermas's theory of communicative action. Scand. J. Manag. **4**, 19–30 (1988)
5. Froomkin, A.M.: Habermas@discourse.net: toward a critical theory of cyberspace. Harv. Law Rev. **116**, 751–873 (2003)
6. Umapathy, K.: An investigation of W3C standardization processes using rational discourse. In: AIS Special Interest Group on Pragmatist IS Research (SIGPrag) Meeting (2009)
7. Habermas, J.: The Theory of Communicative Action: Reason and the Rationalization of Society. Beacon Press, Boston (1984)
8. Lehr, W.: Standardization: understanding the process. J. Am. Soc. Inf. Sci. **43**, 550–555 (1992)
9. Zhao, K., et al.: Vertical e-Business standards and standards developing organizations: a conceptual framework. Electron. Mark. **15**, 289–300 (2005)
10. Choi, B., et al.: Process model for e-Business standards development: a case of ebXML standards. J. Am. Soc. Inf. Sci. **56**, 448–467 (2009)
11. Heng, M.S.H., De Moor, A.: From Habermas's communicative theory to practice on the internet. Inf. Syst. J. **13**, 331–352 (2003)
12. Bolton, R.: The importance of Habermas's "Communicative Action" for social capital and social network theory. In: Annual Meeting of Association of American Geographers. Association of American Geographers (2005)
13. Ngwenyama, O.K., Lee, A.S.: Communication richness in electronic mail: critical social theory and the contextuality of meaning. MIS Q. **21**, 145–167 (1997)
14. Hansen, S., et al.: Wikipedia as rational discourse: an illustration of the emancipatory potential of information systems. In: Hawaii International Conference on Systems Science (HICSS) (2007)
15. Cecez-Kecmanovic, D., Janson, M.: Communicative action theory: an approach to understanding the application of information systems. In: Australasian Conference on Information Systems (1999)
16. Klein, H.K., Huynh, M.Q.: The Critical social theory of Jürgen Habermas and its implications for IS research. In: Mingers, J., Willcocks, L. (eds.) Social Theory and Philosophy for Information Systems, pp. 157–237. Wiley, Chichester (2004)
17. Hansen, S., et al.: Wikipedia, critical social theory, and the possibility of rational discourse. Inf. Soc. **25**, 38–59 (2009)
18. Manninen, T.: Interaction forms and communicative actions in multiplayer games. Int. J. Comput. Game Res. **3**, 5–10 (2003)
19. Yin, R.K.: Case Study Research: Design and Methods. Sage Publications, Thousand Oaks (2003)
20. W3C-XMLWG: W3C. http://www.w3.org/2000/xp/Group/
21. SOAP: W3C. http://www.w3.org/TR/soap12-part1/
22. Krippendorff, K.: Content Analysis: An Introduction to Its Methodology. Sage Publications, Inc., Thousand Oaks (2003)
23. Stemler, S.: An overview of content analysis. Res. Eval. **7**, 137–146 (2001)
24. ATLAS.ti: ATLAS.ti Scientific Software Development GmbH, Berlin, Germany (2015)

25. Rourke, L., et al.: Methodological issues in the content analysis of computer conference transcripts. IJAIED **12**, 8–22 (2001)
26. Neuendorf, K.A.: The Content Analysis Guidebook. Sage Publications, Inc., Thousand Oaks (2001)
27. Lombard, M., et al.: Content analysis in mass communication: assessment and reporting of intercoder reliability. Hum. Commun. Res. **28**, 587–604 (2002)

Acquiring High Quality Customer Data with Low Cost

Xiaoping Liu and Xiao-Bai Li[✉]

Department of Operations and Information Systems,
Manning School of Business, University of Massachusetts Lowell,
Lowell, MA 01854, USA
xiaoping_liu@student.uml.edu, xiaobai_li@uml.edu

Abstract. This work concerns optimal customer data acquisition and selection problem. We first propose using a divide-and-conquer technique to find empirical distribution of the customer data. We then formulate a data acquisition problem as an optimization problem that maximizes the quality of the acquired data while keeping the cost of acquisition as low as possible. We propose using generalized second-price (GSP) auction to acquire customer data and show that when the number of bidders is large, GSP is a truth-telling mechanism. We derive the analytical solution for the optimization problem, which finds a set of data that best represents the probability distribution of the target population relative to the acquisition cost. An experimental study is conducted to demonstrate the effectiveness of the proposed approach.

Keywords: Data acquisition · Data selection · Auctions · Optimization

1 Introduction

In today's data-driven business environment, customer data is the lifeblood of business. Organizations rely on data to perform business analytics to improve operations, gain managerial insights, and develop business strategy in order to gain competitive advantages. Organizations often already have a large amount of customer data in their databases. In many applications, however, it is necessary to acquire new data for business analytics. The two scenarios below demonstrate two different customer data acquisition tasks for different business applications.

Scenario 1. A retail company considers whether or not to carry an additional product line. The product line is new to the company, but related products exist on the market. The company has a database of its customers, but has no data for the new product line. If the company can acquire its customers' purchase data for the related products from other market sources, then the company will be able to make a reasonable estimate about the sales of these new products. In this scenario, the objective is to obtain *aggregated information* about some pre-specified new attributes (e.g., average purchase amount per customer for the new products).

© Springer International Publishing Switzerland 2016
V. Sugumaran et al. (Eds.): WEB 2015, LNBIP 258, pp. 54–65, 2016.
DOI: 10.1007/978-3-319-45408-5_5

Scenario 2. An online company plans to develop an analytic model to predict the amount of purchase on some of its products that its customers buy from other channels (either online or in-store). The model can be used to identify customers who have under purchased the products from the company so that the company can focus its marketing effort on right customers. In this scenario, the objective is to develop a *predictive model* for a new target variable (purchase amount from other channels) based on the data acquired. Alternatively, the objective can also be to acquire data that will be used as predictor variables (rather than the target variable) that are not available in the current customer databases.

A company may have millions of customer records. It is practically impossible to acquire the data from all of its customers. For the purposes of information gathering and business analytics, it suffices to acquire the data from a representative sample of the customer population. Table 1 provides an illustrative example of a sample set, where the left panel shows the customer data currently available and the right panel shows the acquired data for Scenarios 1 and 2, respectively. For convenience, we list the acquired

Table 1. An illustrative example.

ID	Age	Zip Code	Category-1 Purchase	Category-2 Purchase	Category-? Purchase	Scenario 1: Purchase on New Products	Scenario 2: Purchase from Other Channels
1	25	25001	$120	$540	...	$0	$550
2	31	22002	$130	$550	...	$0	$0
3	32	29003	$150	$600	...	$0	$780
4	36	18004	$100	$490	...	$250	$0
5	43	24005	$170	$650	...	$0	$230
6	48	32006	$200	$700	...	$320	$990
7	50	20007	$160	$570	...	$400	$0
8	53	29008	$230	$730	...	$360	$890
9	56	23009	$210	$620	...	$490	$140

data in the two scenarios for each customer in the sample set, although in practice each scenario may correspond to different customer samples.

A straightforward way to get a sample set is to apply a simple random sampling from the customer population. The selected individuals are then solicited to provide the data. There are two serious limitations with this approach [6]. First, response rate is typically too low, particularly when customer data requested include sensitive items. Second, those who respond may be different from those who do not, in terms of demographic and geographic profiles, as well as purchase patterns. Such a sample is not a good representative of the customer population.

Customer data may also be purchased from data brokers or vendors such as Acxiom and Datalogix. However, data brokers rarely have business specific data. For example, the product purchase data shown in Scenarios 1 and 2 in Table 1 are unlikely to be found from a data broker's databases. If the purpose of purchasing data from data brokers is to get additional information for the current customers, it may not be easy to correctly match an individual in the purchased data with a current customer.

This study considers purchasing additional data directly from the current customers. A simple way to do this is to set a fixed price to pay all customers who provide the requested data. However, individuals have different valuations of their personal information. For a fixed price, there may be too many responses from college students and too few from wealthy customers. As a result, the set of respondents does not represent well the customer population. A better way to draw customers with different valuations of their data is to use a procurement auction; that is, to let customers specify the price of their data and then pay the customers based on some payment criteria. If the payment is based entirely on the specified price, then we will likely still have a biased sample set, for example, with too many college students and too few wealthy customers. To avoid this problem, data acquisition needs to consider not only the cost of the data, but also the quality of the data in terms of how well the acquired data represents the customer population.

In this study, we propose a novel approach for customer data acquisition. The proposed approach first applies a divide-and-conquer technique called kd-trees to find the empirical distribution of the customer data, which is represented by a set of *empirical distribution cells* (EDCs) that contains relatively homogeneous customer records. A generalized second-price (GSP) auction is then conducted for each EDC to obtain customers' bid prices for the requested data. We formulate the data acquisition problem as an optimization problem that maximizes the quality of the acquired data while keeping the cost of acquisition as low as possible. We derive the analytical solutions for the optimization problem, which are used to determine the amount of data to purchase.

The main contributions of this paper include: (1) We propose using kd-trees to find empirical distribution of the customer data and representing the distribution by a set of EDCs. (2) We propose using GSP auction to purchase customer data and show that when the number of bidders is large, GSP approaches a truth-telling mechanism. (3) We consider data purchase and data usage simultaneously and formulate the data acquisition problem as an optimization problem to maximize data quality while keeping acquisition cost as low as possible. Furthermore, we derive closed form solutions for the optimization problem.

2   Literature Review

Early studies on information acquisition have been conducted by Moore and Whinston [10, 11], which develop optimal information-gathering strategy for decision model building. The works are later generalized by Mookerjee and Santos [8] and Mookerjee and Mannino [9] to develop optimal data acquisition approaches for decision support systems.

Data acquisition problem has also been studied in the area of active learning. In particular, Saar-Tsechansky et al. [12] propose an active feature acquisition approach for classification applications. Their study considers acquiring instances where some attribute values are missing. Their method uses an instance's values of non-missing attributes to evaluate its contribution to the performance of a classification model. An instance will be acquired if it can significantly reduce the classification errors on a per unit cost basis. After the instances are acquired, they are added to the original data to form a new dataset to build a classification model.

In the afore-mentioned studies, the data are acquired for existing predictor/input attributes or target/outcome attributes. The selection of data depends largely on their impact on the performance of the prediction or decision models. A common limitation of these approaches is that they cannot be used to select or acquire data for a new attribute. So, these approaches cannot be applied to the two data-acquisition problem scenarios described earlier in this paper.

Zheng and Padmanabhan [14] study a data acquisition problem that involves acquiring new predictor (input) attributes. Their approach selects the data based on a prediction model built on existing data. In their work, two scores are computed for each candidate instance: one score measures the instance's contribution to imputation of unknown attribute values; the other score measures the instance's contribution to prediction accuracy. The weighted value of the two scores is then used to select instances. The purpose of data acquisition is to build prediction models. Their method is designed to acquire predictor (input) attributes. It cannot be used to acquire data for a new attribute not yet related to a prediction model, such as Scenario 1 above, or a new target (outcome) attribute, such as Scenario 2 above.

The above data-acquisition studies are all related to some decision models. The essential objective of these model-based approaches is to improve model performance. Consequently, selected data points are more likely to be located in areas that are highly critical to the model (i.e., near the model's decision boundary). These techniques are obviously not applicable to data acquisition problems that do not involve building prediction models.

For data acquisition problems that are not related to model building, a usual approach for data selection is random sampling. We have discussed earlier two limitations of simple random sampling; i.e., low response and biased sample characteristics. Simple random sampling also is not a good choice for purchasing data directly from customers because it can be very expensive. For example, some customers asking very high price for their data may be selected even though there exist many customers with similar profiles and lower prices (but are not selected due to random sampling). A better alternative is to use stratified sampling, where the population is first divided into groups based on some attributes (e.g., age and location) and a sample of appropriate size is then selected within each group. However, customer data usually have a large number of numeric and categorical attributes. There is not a well-accepted approach in the literature to perform stratified sampling on such high-dimensional data [2].

Furthermore, none of the above-mentioned data-acquisition studies consider how to acquire data and how to determine the cost of the data. These studies assume that data acquisition procedure is known and the cost of the data is given. Some of the studies deal with the initial data acquiring phase, which is related to the cost of the data, and final data selection phase, which is related to the quality of the data, in a separate manner [10, 11]. Our work considers the cost and quality of the data simultaneously, and formulates and solves the data acquisition problem in an integrated manner.

3 The Proposed Data Acquisition Approach

3.1 Finding Empirical Distribution

In order for the acquired dataset to be a good representative of the customer population, it is important to effectively find empirical distribution of the customer population (we assume that customer population cannot be characterized by a known probability distribution due to its high-dimensions and complex data types). We propose using kd-trees [4] to find empirical distribution of the customer data. A kd-tree is a divide-and-conquer technique for partitioning data into homogeneous subsets. It recursively divides a dataset into smaller subsets such that data points within each subset are more homogeneous after each partition. For numeric or ordered data, a kd-tree algorithm typically selects the attribute with the largest normalized variance and splits the data into two subsets at the median or mid-range of the attribute. After each partition, the attribute values within a subset are relatively closer to each other. Most categorical data can be ordered and then treated as numeric data. If not, categorical data will be converted to binary data with 0–1 coding. For a dataset of N records, the computational time complexity for kd-trees is of $O(N \log N)$, which is very efficient.

Our idea is to use a kd-tree to partition the customer population data into a number of subsets, within which customer records are relatively homogeneous. Each subset forms an *empirical distribution cell* (EDC) and the collection of all EDCs represents the empirical distribution of the customer population (in the same sense that a histogram represents a univariate distribution). Since the size of the customer population is large, the size of each EDC can also be fairly large. Data selection can then be performed within each EDC separately.

To see how a kd-tree works, consider the dataset in Table 1. The data for the first three attributes (Age, Zip Code and Category-1 Purchase) are plotted in Fig. 1, where the size or labeled value of a bubble represents a Category-1 purchase amount. The algorithm first finds the attribute with the largest normalized variance, which is Age, and splits the data at the mid-range of Age, which is 40.5. Next, within each subset, Zip Code has the maximum variance and is then selected to split the data. The dataset is now divided into four subsets, each containing data points that are more homogeneous in Age and Zip Code (and perhaps in Category-1 Purchase as well). Suppose the original dataset has millions of records. The dataset will be partitioned recursively until the ranges of the attribute values for the subsets are smaller than the pre-specified threshold values. Each subset then becomes an EDC, which can still be fairly large with, say, hundreds of records.

The kd-tree based partitioning is appropriate when the purpose of data acquisition is for gathering additional attribute information (e.g., Scenario 1) rather than building predictive models. If data acquisition is tied to a prediction model (e.g., Scenario 2), the classification and regression trees can be used to partition the data and get EDCs.

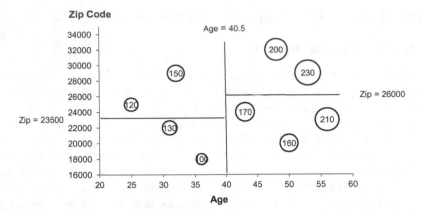

Fig. 1. Example data partitioned by kd-tree.

3.2 Data Acquisition Using GSP Auction

To attract a sufficient number of participants, the auction used for purchasing data should be able to result in multiple winners. It is also important to use a truth-telling mechanism for data acquisition, where there is no incentive for a bidder to submit a price that deviates from the bidder's true valuation of the data. A well-known truth-telling multiple-winner auction is the classic Vickrey-Clarke-Groves (VCG) auction. However, computation of VCG prices is convoluted and hard for general consumers to understand; so, it is impractical to implement the VCG auction in our problem setting. Another recognized mechanism is the Becker-DeGroot-Marschak (BDM) auction, where the payment depends on a randomly drawn number. However, we can show that *a BDM auction incurs higher acquisition cost than a GSP auction for the same customer data* (the detail of this result is not discussed in this paper due to space limitations). As a result, we propose using the GSP auction for data acquisition.

The GSP auction has been used by Google (AdWords) and some other search engines to run keyword advertisements online. Advertisers submit bids for some keywords to a search engine, together with sponsored links related to the keywords. When a keyword matches a query of a search engine user, the search engine will show, along with the normal unpaid search result, a list of matching sponsored links. The listing order of the sponsored links depends on the rankings of the bids for the keyword. If a link is clicked, the advertiser will pay to the search engine a fee calculated based on the advertiser's bid for the keyword. Specifically, the advertiser with the highest bid pays an amount equal to the second highest bid, the advertiser with the second highest bid pays an amount equal to the third highest bid, and so on.

Practically, GSP is easy to implement. It can be used to award a large number of bidders at one go. In our problem context, the bidders participate in an auction to sell rather than to buy. So, it is a reverse GSP auction. The winners are selected by their bids from the lowest to the highest. The person with the lowest bid is paid an amount equal to the second lowest bid, the person with the second lowest bid is paid an amount equal to the third lowest bid, and so on. The GSP auction is not an exactly truth-telling mechanism [3, 13]. However, as stated in Theorem 1 below, GSP approaches a truthful mechanism with a large sample.

Theorem 1. *When the number of bidders is large, GSP is asymptotically a truth-telling mechanism.*

The proofs of Theorem 1 and all other mathematical results are provided in the Appendix. Since our application involves many bidders, GSP becomes a very appealing auction mechanism.

3.3   Data Acquisition Model

Let m be the number of EDCs. In our proposed data acquisition scheme, there will be a total of m auctions, each applied to an EDC. Let x_i $(i = 1,\ldots,m)$ be the number of records acquired in the ith EDC. To facilitate analysis, we assume that bidders' values follow a uniform distribution. The uniform distribution assumption is very common in auction theory literature [7]. For the ith EDC, let d_i be the average difference in price between two consecutive bids. Without loss of generality, set the lowest bid to zero. With the uniform distribution, the *expected cost* for buying x_i records is $\frac{1}{2}d_i x_i^2$. So, the total cost (payment to the bidders) is $\frac{1}{2}\sum_{i=1}^{m} d_i x_i^2$.

The quality of the data selected is measured by the closeness between the distribution of the selected data and that of the population. We note from the context of Scenarios 1 and 2 that the population refers to the *company's customer data collection* (but not a general consumer population). A classical measure of the closeness of the actual and expected distributions is the chi-squared goodness-of-fit statistic, $X^2 = \sum_i \frac{(o_i - e_i)^2}{e_i}$, where o_i is the frequency count based on the observed distribution and e_i is the frequency count on the expected distribution. Let n be the total number of records to be acquired and p_i be the proportion of the records in the ith EDC from the population. Then, the chi-squared statistic in our context can be written as $X^2 = \sum_{i=1}^{m} \frac{(x_i - np_i)^2}{np_i}$. Clearly, the smaller the X^2 value, the closer the two distributions.

Our data acquisition problem has two objectives: (1) to best represent the population with the acquired data, and (2) to minimize the cost for acquiring data. To balance the tradeoff between the two objectives, the objective function of our problem is expressed as a weighted combination of the two objectives. The data acquisition problem can then be formulated as follows:

$$\min \quad \sum_{i=1}^{m} \frac{(x_i - np_i)^2}{np_i} + \frac{1}{2}w \sum_{i=1}^{m} d_i x_i^2, \tag{1.1}$$

$$\text{s.t.} \quad x_i \geq 1, \, i = 1,\ldots,m, \tag{1.2}$$

$$\sum_{i=1}^{m} x_i = n,, \tag{1.3}$$

where the weight w should be specified by considering the scale of the two terms in (1.1). The decision variable of this problem is x_i $(i = 1, \ldots, m)$. Constraint (1.2) ensures that each EDC has at least one record selected. Constraint (1.3) requires that the total number of records taken from all EDCs should sum up to n. This problem can be solved analytically with the following results:

Theorem 2. *The optimal solution for problem* (1) *is*

$$x_i = \frac{np_i}{2 + wnp_id_i} \bigg/ \sum_{j=1}^{m} \frac{p_j}{2 + wnp_jd_j}, \quad i = 1, \ldots, m. \tag{2}$$

Note that the solution for x_i should be rounded to an integer, as commonly practiced in sample size determination.

Example. Consider a population consisting of two EDCs $(m = 2)$, characterized by college students (EDC1) and non-students (EDC2). The population distribution is $p_1 = 0.3$ and $p_2 = 0.7$. Suppose we plan to acquire $n = 100$ individual records from this population. Using GSP, there are 600 bids from college students and 400 bids from non-students. The bid prices are evenly distributed over [\$0.02, \$12] for students and [\$0.05, \$20] for non-students. So, we have $d_1 = 12/600 = \$0.02$ and $d_2 = 20/400 = \$0.05$. Let $w = 0.5$. Substituting these values into (2), we have $x_1 = 41$ students and $x_2 = 59$ non-students. The cost for x_1 is $\$0.02 \times 41^2/2 = \$16.81$ and for x_2 is $\$0.05 \times 59^2/2 = \$87.03$. So, the total cost with GSP is $\$16.81 + \$87.03 = \$103.84$.

If a paid random sampling is used, we can expect that a similar proportion (600:400) of students and non-students are interested. Then the selected sample set will include about $x_1 = 60$ students and about $x_2 = 40$ non-students. Clearly, the student/non-student distribution with GSP (41:59) is much closer to the population distribution (30:70) than that with random sampling (60:40). Since the average cost is $\$12/2 = \$6$ for a college student and $\$20/2 = \$10$ for a non-student, the expected total cost with random sampling will be $6(60) + 10(40) = \$760$, which is much more expensive than the cost with GSP (\$103.84).

4   Experiments

To evaluate the proposed data acquisition approach, we performed an experiment on real data. The dataset, extracted from the US Census Bureau's data website [1], includes financial data for 1,080 individuals. There are 9 numeric attributes, including various categories of income, interests, taxes, and deductions. The dataset was divided into 12 EDCs using a kd-tree. Random sampling and GSP were then used to select a sample of 100 records. The prices were generated uniformly distributed in [\$1, \$1000]. The weight parameter w was set to 0.5 for GSP.

The frequency distributions with random sampling and GSP are shown in Fig. 2, along with the population's distribution. It is clear that the distribution with GSP is much closer to that of the population than the distribution with random sampling, indicating that the data obtained by GSP have higher quality than those by random sampling. Table 2 shows the number and cost of selected records for random sampling and GSP. We can see that the number of records selected with GSP for each EDC is equal to the expected number calculated based on the relative size of the EDC. The number of records selected with random sampling, however, deviates considerably from the expected number. On the other hand, the cost with GSP ($1,153) is significantly lower than the cost with random sampling ($6,260).

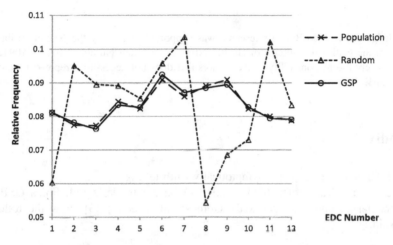

Fig. 2. Comparisons of frequency distributions.

Table 2. Comparison of acquisition costs and sample sizes

EDC	# records in EDC	Expected # records in EDC	# Records with rand. sampling	Costs of rand. sampling	# records with GSP	Costs of GSP
1	88	8	6	375.81	8	93.56
2	84	8	10	661.85	8	101.74
3	83	8	9	485.79	8	75.62
4	91	8	9	487.19	8	83.41
5	89	8	9	555.43	8	100.09
6	98	9	10	678.24	9	123.28
7	93	9	10	727.02	9	114.78
8	96	9	5	310.27	9	92.97
9	98	9	7	356.89	9	84.77
10	89	8	7	483.69	8	102.07
11	86	8	10	593.98	8	84.93
12	85	8	8	544.29	8	95.95
Total	1080	100	100	$6260.40	100	$1153.20

5 Conclusion

In this study, we propose a new data acquisition approach that maximizes the quality of the acquired data while keeping the cost of acquisition as low as possible. We have shown how to find a set of data that best represents the probability distribution of the target population relative to the acquisition cost. The work described so far focuses on situations where the purpose of data acquisition is for gathering additional attribute information (e.g., Scenario 1). We have also been working on data acquisition problems related to prediction models (e.g., Scenario 2) and will report the results in the near future.

Acknowledgments. Xiao-Bai Li's research was supported in part by the National Library of Medicine of the National Institutes of Health (NIH) under Grant Number R01LM010942. The content is solely the responsibility of the authors and does not necessarily represent the official views of NIH.

Appendix

Proof of Theorem 1 (GSP is asymptotically truth-telling).
Let v be the true value of a bidder and $\beta(v)$ be the bidder's bid. For a GSP with N bidders (large) and S slots (fixed), Gomes and Sweeney [5] show the following relationship:

$$\beta(v) = v - \sum_{s=2}^{S} \gamma_s(v) \int_0^v (v - \beta(x))F^{N-s-1}(x)f(x)dx,$$

where F (with the density f) represents the distribution of bidders' values and $\gamma_s(v)$ is a function of v with a complicated expression. We have derived the result $\beta(v) = v$ by showing that

$$\lim_{N\to\infty} \sum_{s=2}^{S} \gamma_s(v) \int_0^v (v - \beta(x))F^{N-s-1}(x)f(x)dx = 0.$$

The details are omitted due to space limitations. Instead, we provide a sketch of the proof here in a Nash equilibrium context. First, suppose that the bidder underbid with $\beta^-(v) < v$ and wins. He will be paid an amount next to his price $\beta* > \beta^-(v)$. If $\beta* > v$, then the bidder would win with bid v anyway. If $\beta^* \le v$, then the payoff to the bidder would be $\beta^* - v \le 0$. So, underbidding does not improve the bidder's payoff. Now, suppose that the bidder overbid with $\beta^+(v) > v$. Since N is large and S is fixed, there will be a large number of bids within $[v, \beta^+(v)]$ and this number will be larger than S. Consequently, the bidder will not win by overbidding.

Proof of Theorem 2 (Solutions to the optimization problem (1)).
We standardize the minimization problem (1) to

$$\min \sum_{i=1}^{m} \frac{(x_i - np_i)^2}{np_i} + \frac{1}{2} w \sum_{i=1}^{m} d_i x_i^2, \tag{A.1}$$

$$\text{s.t.} \, 1 - x_i \leq 0, \, i = 1, \ldots, m, \tag{A.2}$$

$$\sum_{i=1}^{m} x_i - n = 0. \tag{A.3}$$

Applying the KKT conditions to this problem, we have for Lagrange multipliers $\lambda_i, i = 1, \ldots, m,$

$$\lambda_i \geq 0, \, i = 1, \ldots, m,$$
$$\lambda_i(-x_i + 1) = 0, \, i = 1, \ldots, m,$$
$$\sum_{i=1}^{m} x_i - n = 0. \tag{A.4}$$

It follows from (A.1), (A.2) and (A.3) that

$$\frac{d}{dx_i} \left(\sum_{i=1}^{m} \frac{(x_i - np_i)^2}{np_i} + \frac{1}{2} w \sum_{i=1}^{m} d_i x_i^2 + \sum_{i=1}^{m} \lambda_i(-x_i + 1) + \mu(\sum_{i=1}^{m} x_i - n) \right) = 0, \, i = 1, \ldots, m,$$

or

$$\frac{2(x_i - np_i)}{np_i} + wd_i x_i - \lambda_i + \mu = 0, \, i = 1, \ldots, m.$$

For non-boundary solutions, $x_i > 1$. It then follows from (A.4) that $\lambda_i = 0, i = 1, \ldots, m$. Thus,

$$2(x_i - np_i) + wnp_i d_i x_i + np_i \mu = 0, \, i = 1, \ldots, m,$$
$$x_i(2 + wnp_i d_i) - 2np_i + np_i \mu = 0, \, i = 1, \ldots, m,$$

i.e.,

$$x_i - \frac{2np_i}{(2 + wnp_i d_i)} + \mu \frac{np_i}{(2 + wnp_i d_i)} = 0, \, i = 1, \ldots, m, \tag{A.5}$$

or

$$\sum_{i=1}^{m} x_i - \sum_{i=1}^{m} \frac{2np_i}{(2 + wnp_i d_i)} + \mu \sum_{i=1}^{m} \frac{np_i}{(2 + wnp_i d_i)} = 0.$$

Substituting (A.3) into the above equation, we have

$$n - \sum_{i=1}^{m} \frac{2np_i}{(2 + wnp_i d_i)} + \mu \sum_{i=1}^{m} \frac{np_i}{(2 + wnp_i d_i)} = 0.$$

Solving it for μ, we get $\mu = 2 - \dfrac{1}{\sum_{i=1}^{m} \frac{p_i}{(2 + wnp_i d_i)}}$. Substituting it into (A.5), we have

$$x_i = \frac{np_i}{2 + wnp_i d_i} \bigg/ \sum_{j=1}^{m} \frac{p_j}{2 + wnp_j d_j}, \ i = 1, \ldots, m.$$

This solution satisfies KKT condition and the objective function is convex, so it is the unique solution.

References

1. Brand, R., Domingo-Ferrer, J., Mateo-Sanz, J.M.: Reference data sets to test and compare SDC methods for protection of numerical microdata. European Project IST-2000-25069 CASC (2002)
2. Chaudhuri, A., Stenger, H.: Survey Sampling: Theory and Methods. Marcel Dekker Inc., Florence (1992)
3. Edelman, B., Ostrovsky, M., Schwarz, M.: Internet advertising and the generalized second-price auction: selling billions of dollars worth of keywords. Am. Econ. Rev. **97**(1), 242–259 (2007)
4. Friedman, J.H., Bentley, J.L., Finkel, R.A.: An algorithm for finding best matches in logarithmic expected time. ACM Trans. Math. Softw. **3**(3), 209–226 (1977)
5. Gomes, R., Sweeney, S.: Bayes-Nash equilibria of the generalized second price auction. http://ssrn.com/abstract=1429585
6. Groves, R.M.: Nonresponse rates and nonresponse bias in household surveys. Public Opin. Q. **70**(5), 646–675 (2006)
7. Krishna, V.: Auction Theory. Elsevier, Amsterdam (2010)
8. Mookerjee, V., Dos Santos, B.: Inductive expert system design: maximizing system value. Inf. Syst. Res. **4**(4), 111–131 (1993)
9. Mookerjee, V., Mannino, M.: Redesigning case retrieval to reduce information acquisition costs. Inf. Syst. Res. **8**(1), 51–69 (1997)
10. Moore, J., Whinston, A.: A model of decision-making with sequential information-acquisition (Part 1). Decis. Support Syst. **2**(4), 285–307 (1986)
11. Moore, J., Whinston, A.: A model of decision-making with sequential information-acquisition (Part 2). Decis. Support Syst. **3**(1), 47–73 (1987)
12. Saar-Tsechansky, M., Melville, P., Provost, F.: Active feature-value acquisition. Manage. Sci. **55**(4), 664–684 (2009)
13. Varian, H.R.: Position auctions. Int. J. Ind. Organ. **25**(6), 1163–1178 (2007)
14. Zheng, Z., Padmanabhan, B.: Selectively acquiring customer information: a new data acquisition problem and an active learning-based solution. Manag. Sci. **52**(5), 697–712 (2003)

The Mobile Internet as Antecedent for Down-Scoping Corporate Service Portfolios

Claudia Loebbecke[1(✉)], Virpi K. Tuunainen[2], and Stefan Cremer[1]

[1] University of Cologne, Cologne, Germany
{claudia.loebbecke,stefan.cremer}@uni-koeln.de
[2] Aalto University, Helsinki, Finland
virpi.tuunainen@aalto.fi

Abstract. This paper investigates how technological innovations such as broadband mobile Internet and the development to performance- and search-based online advertising determine the corporate scope of established Internet players. Theoretically grounded in the resource-based view and the theory of dynamic capabilities and complemented by a contingency perspective, this study investigates an established Internet player offering domain trading and domain parking services. It finds that the widely appreciated, exogenous technological innovations of the mobile Internet act as antecedent of change in corporate scope and service offerings, which cause not only up-scoping, but also down-scoping decisions [1]. Thereby, the paper contributes to the literature on resource- and capability based opportunities in technologically changing e-business environments and complements prior research on exogenously driven corporate scope and e-business model developments.

Keywords: e-business · Service portfolio · Contingency perspective · Resource-based view · Corporate scope · Domain trading and parking

1 Introduction

The trend of the Internet going mobile – fostered by the growing penetration of mobile broadband access, lower data plan costs and the diffusion of smart-phones – has been widely acknowledged. Early mobile telephones, so-called feature phones, have been important access devices for about two decades. They have been replaced by a second generation of devices, including netbooks, tablets, and smartphones. The number of mobile Internet users is growing and has attracted – among others – interest of the industry for direct products offerings on mobile devices and mobile advertising [2–4].

In many settings, the technological innovations accompanying the Internet going mobile act as antecedents of change with regard to corporate scope and business models in the sense that technological innovations prompt companies to reconfigure their corporate service portfolios [5]. Prior research has covered the impact of a company's resources and innovation capabilities in determining its corporate scope and business model for service offerings [6, 7]. With this research, we explore the impact of exogenous technological innovations on the same.

© Springer International Publishing Switzerland 2016
V. Sugumaran et al. (Eds.): WEB 2015, LNBIP 258, pp. 66–77, 2016.
DOI: 10.1007/978-3-319-45408-5_6

We investigate how the technological innovations such as the development of the mobile Internet from narrow-band accessed via feature phones to broad-band accessed via smart-phones and changes in online advertising such as the growth of search-based and performance-based advertising contribute to determining a company's corporate scope and business model for service offerings. Taking a contingency perspective, we examine how those exogenous technological innovations will be followed by a reconfiguration of established Internet players' service portfolio. In particular, in the context of a globally leading provider of niche Internet services, we pursue the following *research question* with this qualitative work: *how do the technological innovations of the mobile Internet promote down-scoping a profitable corporate service portfolio over up-scoping it?* In other words, different from related literature on business models, the central contribution of this paper lies in examining how highly praised technological innovations not only cause established players to expand their activity scope, but also to down-scope and divest.

To this end, this paper summarizes the theoretical grounds of the resource-based view and dynamic capabilities complemented by the perspective of contingency theory, here standing for specific exogenous technological innovations, and then applies those theoretical grounds to the case of *Domain-Gain*[1], a leading established Internet player in the niche domain trading and domain parking services. Building on the academic discourse on resources and capabilities determining a company's scope and business model, the paper discusses the findings with regard to the role of exogenous technological innovations, here the development of the mobile Internet and online advertising, as impetus for up- and down-scoping a corporate service portfolio.

The remainder of the paper is organized as follows. In the next section, we define relevant terms and describe technology-induced changes in mobile business. We then lay the theoretical grounds of our study by integrating the resource-based view [7–9] and dynamic capabilities [10, 11] with a contingency perspective [12, 13] for investigating external technological innovations as antecedents to corporate up-scoping and down-scoping. We discuss the reconfiguration of an established Internet player's service portfolio against the technological developments and the outlined theoretical grounds. We conclude with some theoretical and practical implications and suggestions for further research.

2 Terms and Technology-Induced Changes in Mobile Business

With this section, we aim at a common understanding of several terms, for which we find many diverging definitions in the scientific literature and the press. Instead of making the claim for the appropriate definition, we want to avoid misinterpretations throughout this work.

The *Mobile Internet*, also termed mobile web, wireless Internet, or wireless web, has been defined with respect to the content formatted to mobile devices [14] or – emphasizing the 'always-on' mentality – focusing on hardware-related aspects such as screen

---

[1] Additional data are disclosed for anonymous review, but available for presentation.

size, processing capabilities and input facilities [15, 16]. We combine both views and define the mobile Internet broadly as the possibility to use Internet content and services on handheld devices at any time, accessed for certain purposes via mobile communication networks, making the access device the only difference between the stationary Internet and the mobile one. We exclude access by WiFi as mobile Internet.

While the terms *feature phone* and *smart-phone* are widely used, there is no commonly accepted definition codifying which characteristics a mobile phone requires in order to qualify for one or the other category [17]. The corresponding definitions change over time; they typically refer to differences in hardware capabilities (sheeted processor, additional sensors, touch screens, cameras) or software (extensibility of the system through the installation of software). It is general understanding that that feature phones are early generation mobile devices, typically designed for placing and receiving calls and text messages using a manufacturer's proprietary operating system. Smartphones refer to high-end, multifunctional devices which resemble mini-computers, compared to feature phones. Prominent examples for smartphones are Apple's iPhone and various devices running Android, Microsoft's Windows Phone or, more recently, Windows 10 Mobile. In this paper, we consider all those devices to be smartphones; we exclude laptops, netbooks, tablets, and other connected devices.

The diffusion of smartphones and the related increase in mobile Internet usage change the context of *mobile ads* and *related advertising-based mobile business models* [18]. Mobile ads resemble their stationary pendants [19–21] both include banners, pop-ups, search-based ads, and interactive elements. In line with mobile advertising, two trends have been observed: (1) performance-based advertising complements and possibly substitutes fix-price advertising [22] and (2) mass advertising loses ground against search-based advertising [23].

- *Performance-Based Advertising.* Performance-based advertising points to directly measuring user responses on ads (e.g., [24]). Payments are based upon click-through rates [23] or on conversion rates. For both, click-through rates and conversion rates, the quality of the landing pages is crucial [25, 26]. Landing pages are distinguished along their page relevance, transparency, and navigability. Performance-based advertising not only complements, but increasingly substitutes fix-price advertising.
- *Search-Based Advertising.* The trend towards search based Internet access has changed consumers' online behavior and subsequently the way online advertising is conducted. Search-based advertising has gained ground over mass advertising [23].

3   Theoretical Grounds

We explore our research question, how the technological innovations of the mobile Internet promote down-scoping over up-scoping of a profitable corporate service portfolio, on the grounds of the *resource-based view* and *theory of dynamic capabilities* complemented by a contingency perspective.

In the resource-based view, a company's competitive position results from processing a bundle of human and physical resources [7–9]. Heterogeneously distributed, unique resources, that are valuable, rare, inimitable, and non-substitutable (VRIN) could lead to a lasting, resource-based competitive advantage [27]. Beyond counting on physical resources, companies searching for a resource-based competitive advantage need to focus on the processes and capabilities needed to exploit their resources [28–30]: They must manage and align them and install replicable routines for leveraging them [31]. Such process focus requires dynamic capabilities, i.e., a company's abilities "to integrate, build and reconfigure internal and external competences to address rapidly changing environments" [11, p. 516]. Derived from prior knowledge, dynamic capabilities give direction on how to develop, manage, and deploy the resources [10] and hence to drive innovation. Technological capabilities are one influential form of dynamic capabilities [32]. Developed over time and accumulated with experience, they reflect a company's ability to acquire and apply various technologies. They are an influential driver of innovation and of sustaining competitive advantage [33, 34].

However, technological innovations in the business environment may also challenge a company's resources and technological capabilities and eventually foster either entering or exiting certain business activities [35]. Helfat and Eisenhardt [36] demonstrate the role of technological innovation as an antecedent of changes in corporate scope. Companies adjust their corporate portfolio in response to industry conditions [37–39] or a changing technological environment. Studies explicitly investigating changes in corporate scope with technological innovation as antecedent of down-scoping and divestments are still rare. An exception provides Kaul [5] who finds that technological in-house innovations prompt a company to reconfigure its corporate portfolio.

Further related and more IS/ICT oriented works can be found in the literature on business models. While there is still no consistent definition of a business model [32, 40–42], there seems to be agreement that a business model describes how a company interacts and transacts with its customers and partners and converts resource investments into economic value [43, 44]. Almost every business model conceptualization harks back upon the resource-based view stating that – especially in the digital era – generation of sustainable economic value relies on a company's ability to absorb ICT resources and align them to its other resources [40, 45].

Our interest in this study reaches beyond in-house developed technological innovations to exogenous ones [46–48]. This has encouraged reference to contingency theory [12, 30]. Combining a contingency perspective with the resource-based view [49, 50] allows investigating how externally occurring technological innovations impact the value of organizational resources and associated up-scoping and also down-scoping decisions [1]. This is in line with the more recent ICT-oriented business model literature which consistently finds that in times of the Internet and digital business, companies need to be capable of adapting their business models or their corporate scope to changes in the environment [40, 51–54].

4 The Case of Down-Scoping Domain-Gain's Service Portfolio

Research Methodology and Data Collection. We investigate our research question via an explorative study [55–58] of *Domain-Gain* and its service portfolio including domain trading and domain parking services. For data collection, we use in-depth interviews, server data, and publicly available documents. We conducted sixteen in-depth face-to-face interviews with *Domain-Gain* representatives in three rounds, in the second half of 2010, in the second half of 2013, and in the first half of 2015. Semi-structured interviews – not just following a fixed set of questions – were deemed to be an appropriate method for data collection. Some interrogations took place more than once – during and across interview periods (2010, 2013, and 2015) in order to gather as much information as possible. Whereas all initial interviews were face-to-face at *Domain-Gain's* premises in Europe and the US, some subsequent sessions were based on e-mail, Skype and telephone. We could also access sets of company internal server data – differentiating traffic generated by end-users accessing a website along the country of origin, the web browser used, the generated earnings, and the access date. As *Domain-Gain's* tracking system does not automatically distinguish between stationary and mobile browsers, a time-consuming manual data filtering was unavoidable. Finally, printed and website-based information including publicly accessible reports and press releases allowed us to validate findings derived from other sources.

Domain-Gain's Service Portfolio. The company operates a global marketplace for domain names (URLs) with more than 1.5 million member accounts from around the world. With offices in the United States and Europe, *Domain-Gain* has revenues of about USD 36 million in 2013. For fifteen years, *Domain-Gain* has built its business on two main pillars, domain trading and domain parking, complemented by a portfolio of tools for buying, selling, and monetizing domains.

- *Domain Trading* refers to selling domain names via an online trading platform. *Domain-Gain* offers about 17 million domains for sale and sells more than 2,500 domains monthly. The business of domain trading is based on the scarcity of good domain names in terms of navigation and representation [59]. In 2013, domain trading accounted for about 60 % of *Domain-Gain's* gross revenues and 40 % of its net revenues. Impressive sales figures include fly.com for USD 1.8 million in 2009, mm.com for USD 1.2 million in 2014, and – still beating the rest – sex.com for USD 13 million in 2010.

- *Domain Parking* builds on monetizing unused but parked domains via performance-based advertising [60, 61]. It is based on a domain name naturally generating traffic derived from its unique name. The traffic originates either from typing-in a domain name in the address bar of a web browser or from old back-links coming from prior use of the (expired) domain. Domain owners use *Domain-Gain's* parking service to earn revenue while promoting the domains for sale. They set keywords and domain topics in their *Domain-Gain* account. As soon as a parked domain is accessed on *Domain-Gain's* domain trading platform, *Domain-Gain* places pay-per-click text-link ads and a design-template to purport that it is a real website. It aims at optimizing keywords and ad placements as to fit the expectations of domain visitors. This is different from sponsored advertising, where advertisers submit their product information in the form of specific keyword listings to search engines bidding for promising listing on the search engine results page. When a user searches for a term on the search engine, the advertisers' webpage appears as a sponsored link next to the organic search results (e.g., [62]). Advertising on parked domains is performance-based; the domain owner receives a per-click revenue share.

Across the markets, *Domain-Gain's* mobile click-through rate has been almost 2.5 times higher than its stationary one. In 2013, about 4 % of the views on *Domain-Gain's* parked domains originating from mobile sources generated 8 % of all ad-clicks and drove about 6 % of Domain-Gain's total gross parking revenue. However, the revenue-per-click (the ad price) was on average lower in the mobile Internet. This limitation of the mobile domain parking business was mostly beyond *Domain-Gain's* control. Google placed more than 90 % of text-ads displayed on parked pages. Hence *Domain-Gain* could only influence the prices for the remaining less than 10 % of ads.

Early on, *Domain-Gain* offered domain parking services complementing its domain trading services. Early on, 'pre-mobile' so-to-say, both business pillars led to a profitable position on a global scale. When the Internet and hence also unused domains went mobile, the expectation was that the stationary position, i.e., resources and capabilities driving stationary success, could be transferred to the mobile world. Mobile domain parking was considered a mere variation of the profitable stationary domain parking business.

This expectation did not fully materialize. The mobile Internet hampered the up-scping of *Domain-Gain's* service portfolio along long four main lines: two concerning the mobile business performance compared to the stationary one and two questioning the general viability of the domain parking business model – stationary and mobile.

- While mobile click-through rates were found to be on average 2.5 times higher than stationary ones, ad prices in the mobile world were lower and thus limited the revenue-per-click.
- With the increasing diffusion of smartphones, users increasingly accessed the Internet via apps which – due to their architecture – did not entail the risk of mistyping web addresses.
- With the shift to performance-based advertising, accidental visitors who are annoyed about landing on a wrong page and hence do not click on the advertisement, fail to contribute to *Domain-Gain's* revenue stream.
- With search-based advertising, a business model built on back-links and direct domain type-ins seems at risk. Significantly less users type or even mistype URLs which is required to get forwarded to pages parked on *Domain-Gain's* servers. In contrast, Twitter users follow real time links.

5   Discussion

We wanted to explore how the technological innovations of the mobile Internet promote down-scoping over up-scoping of a profitable corporate service portfolio.

We found that *Domain-Gain* was committed to adjust its processes in order to extend existing resources and capabilities for maintaining a sustainable competitive advantage when extending its scope and business model to mobile domain parking. By combining its available resources with its competences – both successfully proven in the stationary Internet –, *Domain-Gain* aimed at leveraging synergies between both, domain trading and domain parking and between its stationary and mobile domain parking business in order to gain leadership in the mobile domain parking market [63]. Thus *Domain-Gain* pursued the basic notion of the resource-based view that "it is not the value of an individual resource that matters but rather the synergistic combination or bundle of resources created by the firm" [64, p. 356].

However, we found the limits of profitably managing and realigning resources and capabilities when exogenous technological innovations – such as an increasing number of smartphones accessing the Internet via apps and the growth of performance- and search-based advertising – rescind an established business model or service offerings, here domain parking. Radical changes in the business environment force *Domain-Gain* to reconsider its corporate service portfolio with respect to mobile domain parking.

The resource-based view stresses the importance of unique, valuable, and imperfectly mobile resources and capabilities for achieving competitive advantage [9, 27]. It underlines the need for processes that allow companies to exploit these resources [65, 66]. *Domain-Gain's* resources, for instance the about 17 million domains for trade on its platform, proved valuable in the stationary Internet. Its process of tightly linking those domains 'on sale' with the company's imperfectly mobile advertising resources and capabilities (e.g., targeted ad networks) helped *Domain-Gain* to launch and succeed in domain parking and allowed it to be market leader as long as the business environment was promising. However, given the Internet's ubiquity of service provision and access, which also applies to advertising services, the concept of imperfectly mobile resources loses relevance.

The need to down-scope due to exogenous technological innovations is not specific to the mobile environment [5]. Technology-induced shifts driving performance- and search-based online advertising occur both in the stationary and the mobile Internet. We found that they lead to a potentially shrinking stationary domain parking business and require re-allocating scarce resources between projects and markets. Hence, down-scoping requires having resources at hand which are scalable. In the case of *Domain-Gain*, resource scalability turned out to be challenging with regard to the company's domain parking business which is inevitably interlaced with its global domain trading activities. Country-code top-level domains were about the only scalable resources that could be exploited in the process of down-scoping.

Once the needed resources are available or at least within reach, Teece [10] suggests that a company deploys its dynamic capabilities to innovate for changing business environments and market conditions by reconfiguring the resources [67, 68]. To that end, Teece [10] disaggregates dynamic capabilities into the capacity to sense and shape opportunities and threats, to seize opportunities, and to maintain competitiveness through enhancing, combining, protecting, and – when necessary – reconfiguring the resources. Concerning sensing and shaping opportunities, *Domain-Gain* understood customer needs focusing on device specific and individualized services. By sourcing its mobile traffic by device or at least by operating system, *Domain-Gain* accumulated significant knowledge about mobile Internet users and their viewing and clicking behavior on parked domains and learned that higher mobile click-through rates alone cannot heal the exogenously driven (smartphones, apps, performance- and search-based advertising) lower mobile traffic on parked sites. With regard to seizing opportunities, *Domain-Gain* balanced resources between its core trading business and its high-margin, but risky mobile domain parking. Teece [10] suggests that maintaining evolutionary fitness and escaping from unfavorable routines helps to sustain continuity only until there is a shift in the environment. Accordingly, *Domain-Gain* did not revamp its organization in light of those major shifts in the technological environment or the overall industry context.

Domain-Gain, being confronted with the need to down-scope due to exogenous technological innovations, resembles the critical lines of incumbents who face a new entrant with offering innovative, perhaps even technologically mediocre products or services. Admittedly, the analogy between incumbents facing a new entrant with a disruptive technology may not be one-to-one applicable to *Domain-Gain* being confronted with exogenous technological innovations. However, the lessons to be learnt from the literature on disruptive technologies [69, 70] may explain how new technological innovations act as antecedent for an incumbent's or established player's down-scoping and eventually divestment.

6 Contribution and Further Research

It is widely acknowledged that digitalization and advances in mobile technologies create significant turmoil in many traditional industries, including retail and media. While *Domain-Gain's* core business of domain trading still flourishes, its domain parking model has deteriorated with the Internet going mobile and users' general affinity to use apps and search engines – even though a large part of the Internet remains being accessed via traditional, large screen web browsers. In this context, our explorative study of *Domain-Gain's* domain parking business in the mobile world illustrates an example of the mobile Internet initiating and driving down-scoping a corporate service portfolio based on a business model, which does not fit the changes imposed by exogenous, industry-wide technological innovations [5, 43, 71].

Similar to Kaul [5], our study contributes to the resource-based view by taking a more holistic perspective on the dynamics of – also exogenous – technological innovations and corporate scope. It thereby complements the literature on corporate scope and business model development [40, 72]. In Particular it points out that the widely appreciated, exogenous technological innovations of the mobile Internet act as antecedent of change in corporate scope and service offerings, which cause not only up-scoping, but also down-scoping decisions [1]. Thereby, the paper extend earlier works on resource- and capability-based opportunities in technologically changing e-business environments and complements prior research on exogenously driven corporate scope [46, 48].

Without any doubt, our findings are accidental at best. As with any single explorative study we face the issue of limited generalizability. Further, digging into the different stages of an established niche player launching and potentially exiting the mobile pendant of its successful stationary Internet business, we had to cope with almost a third of the company's life-time and could barely 'control' other contingency factors and changes among interviewees. Finally, securing the publication opportunity for a study with findings that are rather critical for the company under investigation has posed some constraints in terms of data release.

However, we hope that this study can serve as an effective eye-opener and promote further investigations of the often praised technological innovations impacting corporate scope. Intuitively, we consider technological innovations as opportunity for up-scoping product and service offerings [73]. With our study, we wish to have raised awareness the impact of technological innovation on corporate service portfolios is heavily context-dependent. As we have shown for the niche service of domain parking, each player in a value chain or value web requires careful analysis regarding the direction in which technological innovation act as antecedents for up-scoping or down-scoping a corporate service portfolio.

Further research may want to dig deeper into the impetus of other exogenous forces relevant in the context of mobile commerce such as deregulation and globalization. Any such contingency factors not only make research into business model developments [74] increasingly topical, but may also point to innovations as contingency factors causing corporate down-scoping and even divestment [5].

References

1. Aragon-Correa, J., Sharma, S.: A contingent resource-based view of proactive corporate environmental strategy. Acad. Manag. Rev. **28**, 71–88 (2003)
2. Goldman, S.: Transformers. J. Consum. Market. **27**, 469–473 (2010)
3. Scharl, A., Dickinger, A., Murphy, J.: Diffusion and success factors of mobile marketing. Electron. Commer. Res. Appl. **4**, 159–173 (2005)
4. Zwass, V.: Electronic commerce and organizational innovation: aspects and opportunities. Int. J. Electron. Commer. **7**, 7–37 (2003)
5. Kaul, A.: Technology and corporate scope: firm and rival innovation as antecedents of corporate transactions. Strateg. Manag. J. **33**, 347–367 (2012)
6. Hoskisson, R., Hitt, M.: Antecedents and performance outcomes of diversification: a review and critique of theoretical perspectives. J. Manag. **16**, 461–509 (1990)
7. Penrose, E.: The Theory of the Growth of the Firm. Wiley, New York (1959)
8. Barney, J.: Strategic factor markets: expectations, luck, and business strategy. Manag. Sci. **32**, 1231–1241 (1986)
9. Wernerfelt, B.: A resource-based view of the firm. Strateg. Manag. J. **5**, 171–180 (1984)
10. Teece, D.: Explicating dynamic capabilities: the nature and microfoundations of (sustainable) enterprise performance. Strateg. Manag. J. **28**, 1319–1350 (2007)
11. Teece, D., Pisano, G., Shuen, A.: Dynamic capabilities and strategic management. Strateg. Manag. J. **18**, 509–533 (1997)
12. Lawrence, P., Lorsch, J.: Organization and Environment. Harvard Business School, Boston (1967)
13. Thompson, J.: Organizations in Action. McGraw-Hill, New York (1967)
14. Mobile Marketing Association: Mobile Marketing Association Glossary (2008). www.mmaglobal.com/glossary.pdf
15. Chae, M., Kim, J.: What's so different about the mobile internet? Commun. ACM **46**, 240–247 (2003)
16. Chang, Y., Chen, C.: Smart phone-the choice of client platform for mobile comers. Comput. Stand. Interfaces **27**, 329–336 (2005)

17. Pitt, L., Parent, M., Junglas, I., Chan, A., Spyropoulou, S.: Integrating the smartphone into a sound environmental information systems strategy: principles, practices and a research agenda. J. Strateg. Inf. Syst. **20**, 27–37 (2010)
18. Tsang, M., Ho, S., Liang, T.: Consumer attitudes toward mobile advertising: an empirical study. Int. J. Electron. Commer. **8**, 65–78 (2004)
19. Ha, L.: Online advertising research in advertising journals: a review. J. Curr. Issues Res. Advertising **30**, 31–48 (2008)
20. McCoy, S., Everard, A., Polak, P., Galletta, D.: The effects of online advertising. Commun. ACM **50**, 84–88 (2007)
21. Xu, D., Liao, S., Li, Q.: Combining empirical experimentation and modeling techniques: a design research approach for personalized mobile advertising applications. Decis. Support Syst. **44**, 710–724 (2008)
22. Dellarocas, C.: Double marginalization in performance-based advertising: implications and solutions. Manag. Sci. **58**, 1178–1195 (2012)
23. Ghose, A., Yang, S.: An empirical analysis of search engine advertising: sponsored search in electronic markets. Manag. Sci. **55**, 1605–1622 (2009)
24. Hoffman, D., Novak, T.: How to acquire customers on the web. Harvard Bus. Rev. **78**, 179–188 (2000)
25. Gofman, A., Moskowitz, H., Mets, T.: Integrating science into web design: consumer-driven web site optimization. J. Consum. Market. **26**, 286–298 (2009)
26. Goldfarb, A., Tucker, C.: Online display advertising: targeting and obtrusiveness. Market. Sci. **30**, 389–404 (2011)
27. Barney, J.: Firm resources and sustained competitive advantage. J. Manag. **17**, 99–120 (1991)
28. Dehning, B., Stratopoulos, T.: Determinants of sustainable competitive advantage due to IT-enabled strategy. J. Strateg. Inf. Syst. **12**, 7–28 (2003)
29. Melville, N., Kraemer, K., Gurbaxani, V.: Information technology and organizational performance: an integrative model of IT business value. MIS Q. **28**, 283–322 (2004)
30. Ravichandran, T., Lertwongsatien, C.: Effect of information systems resources and capabilities on firm performance. a resource-based perspective. J. Manag. Inf. Syst. **21**, 237–276 (2005)
31. Henderson, R., Cockburn, I.: Measuring competence? Exploring firm effects in pharmaceutical research. Strateg. Manag. J. **15**, 63–84 (1994)
32. Zhou, K., Wu, F.: Technological capability, strategic flexibility, and product innovation. Strateg. Manag. J. **31**, 547–561 (2010)
33. Anderson, P., Tushman, M.: Technological discontinuities and dominant designs: a cyclical model of technological change. Adm. Sci. Q. **35**, 604–633 (1990)
34. Lavie, D., Rosenkopf, L.: Balancing exploration and exploitation in alliance formation. Acad. Manag. J. **49**, 797–818 (2006)
35. Tushman, M., Anderson, P.: Technological discontinuities and organizational environments. Adm. Sci. Q. **31**, 439–465 (1986)
36. Helfat, C., Eisenhardt, K.: Inter-temporal economies of scope, organizational modularity and the dynamics of diversification. Strateg. Manag. J. **25**, 1217–1232 (2004)
37. Anand, J.: Redeployment of corporate resources: a study of acquisition strategies in the U.S. defense industries, 1978–1996. Manag. Decis. Econ. **25**, 383–400 (2004)
38. Anand, J., Singh, H.: Asset redeployment, acquisitions and corporate strategy in declining industries. Strateg. Manag. J. **18**, 99–118 (1997)
39. Levinthal, D., Wu, B.: Opportunity costs and nonscale free capabilities: profit maximization, corporate scope and profit margins. Strateg. Manag. J. **31**, 780–801 (2010)

40. Al-Debei, M., Avison, D.: Developing a unified framework of the business model concept. Eur. J. Inf. Syst. **19**, 359–376 (2010)
41. Osterwalder, A., Pigneur, Y., Tucci, C.: Clarifying business models: origins, present, and future of the concept. Commun. Assoc. Inf. Syst. **16**, 1–25 (2005)
42. Straub, D.: Foundations of Net-Enhanced Organizations. Wiley, New York (2004)
43. Chesbrough, H., Rosenbloom, R.: The role of the business model in capturing value from innovation: evidence from xerox corporation's technology spin-off companies. Indus. Corp. Change **11**, 529–555 (2002)
44. Sharma, S., Gutierrez, J.: An evaluation framework for viable business models for m-commerce in the information technology sector. Electron. Markets **20**, 33–52 (2010)
45. Hedman, J., Kalling, T.: The business model concept: theoretical underpinnings and empirical illustrations. Eur. J. Inf. Syst. **12**, 49–59 (2003)
46. Ford, M.: Rise of the Robots: Technology and the Threat of a Jobless Future. Basic Books, New York (2015)
47. Loebbecke, C., Picot, A.: Reflections on societal and business model transformation arising from digitization and big data analytics: a research agenda. J. Strateg. Inf. Syst. **24**, 149–157 (2015)
48. Rifkin, J.: The Zero Marginal Cost Society: The Internet of Things, the Collaborative Commons, and the Eclipse of Capitalism. Palgrave Macmillan, New York (2014)
49. Datta, D., Guthrie, J., Wright, P.: Human resource management and labor productivity: does industry matter? Acad. Manag. J. **48**, 135–145 (2005)
50. Guthrie, J., Datta, D.: 2008 Dumb and dumber: the impact of downsizing on firm performance as moderated by industry conditions. Organ. Sci. **19**, 108–123 (2010)
51. Brynjolfsson, E., Saunders, A.: Wired for Innovation: How Information Technology is Reshaping the Economy. MIT Press, Cambridge (2010)
52. Casadesus-Masanell, R., Ricart, J.: Competitiveness: business model reconfiguration for innovation and internationalization. Manag. Res. **8**, 123–149 (2010)
53. Demil, B., Lecocq, X.: Business model evolution: in search of dynamic consistency. Long Range Plann. **43**, 227–246 (2010)
54. Markus, M., Loebbecke, C.: Commoditized digital processes and business community platforms: new opportunities and challenges for digital business strategies. Manag. Inf. Syst. Q. **37**, 649–653 (2013)
55. Benbasat, I., Goldstein, D., Mead, M.: The case research strategy in studies of information systems. MIS Q. **11**, 369–386 (1987)
56. Eisenhardt, K., Graebner, M.: Theory building from cases: opportunities and challenges. Acad. Manag. J. **50**, 25–32 (2007)
57. Pratt, M.: From the editors: for the lack of a boilerplate: tips on writing up (and reviewing) qualitative research. Acad. Manag. J. **52**, 856–862 (2009)
58. Yin, R.: Case Study Research: Design and Methods. Sage Publications, Thousand Oaks (2009)
59. Lindenthal, T., Loebbecke, C.: Pricing quality attributes of internet domain names: a hedonic model for words. In: Americas Conference on Information Systems (AMCIS), Savannah, GA (2014)
60. Almishari, M., Yang, X.: Ads-portal domains: identification and measurements. ACM Trans. Web **4**, 1–34 (2010)
61. Loebbecke, C., Weiss, T.: parking in the mobile world: the case of Sedo. In: International Conference on Electronic Commerce, Liverpool (2011)
62. Bradlow, E., Schmittlein, D.: The little engines that could: modeling the performance of world wide web search engines. Market. Sci. **19**, 43–62 (2000)

63. Loebbecke, C., Tuunainen, V.: Extending successful eBusiness models to the mobile internet: the case of Sedo's domain parking. In: Americas Conference on Information Systems (AMCIS), Chicago, IL (2013)
64. Kraaijenbrink, J., Spender, J., Groen, A.: The resource-based view: a review and assessment of its critiques. J. Manag. **36**, 349–372 (2010)
65. Newbert, S.: Empirical research on the resource-based view of the firm: an assessment and suggestions for future research. Strateg. Manag. J. **28**, 121–146 (2007)
66. Rivard, S., Raymond, L., Verreault, D.: Resource-based view and competitive strategy: an integrated model of the contribution of information technology to firm performance. J. Strateg. Inf. Syst. **15**, 29–50 (2006)
67. Galunic, D., Eisenhardt, K.: Architectural innovation and modular corporate forms. Acad. Manag. J. **44**, 1229–1249 (2001)
68. Rindova, V., Kotha, S.: Continuous morphing: competing through dynamic capabilities, form, and function. Acad. Manag. J. **44**, 1263–1280 (2001)
69. Bower, J., Christensen, C.: Disruptive technologies: catching the wave. Harvard Bus. Rev. **73**, 43–53 (1995)
70. Christensen, M., Overdorf, C.: Meeting the challenge of disruptive change. Harvard Bus. Rev. **78**, 67–76 (2000)
71. Cavalcante, S., Kesting, P., Ulhoi, J.: Business model dynamics and innovation: (Re)establishing the missing linkages. Manag. Decis. **49**, 1327–1342 (2011)
72. Zott, C., Amit, R., Massa, L.: The business model: recent developments and future research. J. Manag. **37**, 1019–1042 (2011)
73. Brynjolfsson, E., McAfee, A.: The Second Machine Age: Work, Progress, and Prosperity in a Time of Brilliant Technologies. W.W. Norton, New York (2014)
74. Casadesus-Masanell, R., Ricart, J.: How to design a winning business model. Harvard Bus. Rev. **89**, 100–107 (2011)

Who Is My Audience? Investigating Stakeholder Collapse in Online Social Platforms

Utku Pamuksuz[✉], Sung Won Kim, and Ramanath Subramanyam

College of Business, University of Illinois at Urbana-Champaign,
515 E Gregory Dr., Champaign, IL 61820, USA
{pamuksu2,swk,rsubrama}@illinois.edu

Abstract. Exploring top executive's behavior on social media potentially indicates an important area for information systems research. Engaging in traditional social strategies through digital channels may not always result in strong online images and impressed stakeholders. Traditional perspectives on impression management (IM) have focused on the use of IM on a single targeted audience. In contrast, individuals employing IM tactics in social networking platforms (SNP) must consider multiple audiences simultaneously. We are interested in understanding how top executives leverage SNP for personal gains. Our findings show that when the definition of audience and purposes of IM are re-visited in digital context, some of the traditional lines of thought fall apart.

Keywords: Social networking platforms · Impression management · Top executives · Text mining

1 Introduction

Being active in social networking platforms (SNP) has recently generated an interest among executives [1]. SNP are relatively recent platforms used to influence and create impressions on internal and external stakeholders, which have implications for executive career success. The ease of disseminating information through SNP lets top executives reach target audiences efficiently to create a favorable image in the eye of stakeholders. Our focus is the SNP usage of top executives and understanding their public strategies, and personal gain.

For top executives including CEO's, CFO's, and other executive VP's, influencing the audience inside and outside the organization entails competent impression management [2]. Impression management (IM) is defined as efforts of an individual to generate, defend or otherwise alter an image held by a target audience [3]. Jones and Pittman [6] identified five tactics of IM: ingratiation, supplication, self-promotion, exemplification, and intimidation. Ingratiation involves individuals rendering favors, conforming opinions, or using flattery to gain appreciation from the target audience [4, 5]. Supplication refers to the tactics of an actor who highlights her/his own weakness to invoke empathy from the audience [6]. Self-promotion defines the efforts of an actor who aims to be seen as competent and respected [5]. Exemplification tactics are demonstrations of self-sacrifice for the company or community to

© Springer International Publishing Switzerland 2016
V. Sugumaran et al. (Eds.): WEB 2015, LNBIP 258, pp. 78–82, 2016.
DOI: 10.1007/978-3-319-45408-5_7

portray moral worthiness [5]. Finally, actors use intimidation to present themselves as powerful and dangerous figures that are capable of harming a target [7].

Management scholars have investigated the effects of top executives' IM usage on career success in traditional media [11, 12] and found that IM tactics employed by top executives can lead to additional career achievements [13]. More specifically, Westphal and Stern [19] examined interpersonal influence behaviors of top managers, from an impression management perspective, in obtaining board appointments. Top executives employed IM tactics not only toward board members and peer directors, but also a wider range of audiences including shareholders [14] and journalists [15] to enhance their career success in primarily traditional media platforms.

Top executives, who employ IM tactics in traditional media, often use boundaries between separate audiences to customize their message [8]. Boundaries segregate different audiences in the applied medium. For example, an individual using email can segregate audiences by addressing a specific individual or group. In SNP, the boundaries between audiences become more permeable which leads to sharing information with multiple audiences simultaneously. In contrast to sharing of specifically-selected information with appropriate targeted audience, we observe context collapse which refers to the overlap of an individual's multiple audiences into one single platform [9, 10]. For example, ingratiation towards a specific person might be intended to be private. However, if the actors engage in such behaviors in SNP, intention of the actor will be visible to multiple audiences rather than only the targeted audience. In this situation, actors have more difficulty managing separate impressions across their target audiences in SNP than in offline settings [10]. We question how these publicly visible platforms affect the individual usage of IM tactics because top executives need to manage impressions in front of multiple audiences simultaneously, to achieve their career goals.

In our work, we seek to answer following research questions: Do IM tactics on SNP affect a top executive's career success? Furthermore, which specific IM tactics employed by top executives on SNP are likely to be associated with better career outcomes? Investigation of these questions will contribute to academic and practitioner research on the usage of social networking sites as platforms for enabling career success.

2   Theoretical Development

Our study analyzes the relationships between the dimensions of IM and career success. We draw on Jones and Pittman's IM taxonomy [6], validated by Bolino and Turnley [5], which captures a large domain of IM behaviors and is applicable to both the individual and organizational level of analysis. We investigate the pitfalls and advantages of these tactics as they are employed in the SNP environment. Further, we consider IM tactics as visible not only to targeted audience but also to other unintended observers.

We propose that the employment of IM tactics across a wide audience impacts executive career success. Previous research on this topic has highlighted five IM tactics employed to gain career success [e.g., 16]. In this research, we focus on the same constructs and re-examine the nature of the relationship between these tactics and executive career success in the context of SNP. Our expectation is that when the definition of audience and purposes of IM are re-visited in this context, some of the traditional lines of thought fall apart or need specific re-examination. Specifically, due to the audience collapse; we posit that ingratiation and supplication tactics, when applied in SNP by top executives, are negatively associated with their career success while self-promotion, intimidation, and exemplification maintain their positive associations.

3   Methodology

We present an empirical analysis of impression management tactics in Twitter as a prevalent social networking site. Twitter often serves as a microblogging platform and forms one of the most popular online communities in the world [17]. We use 110 top executives verified twitter accounts listed in S&P 1500 index. We gather all the tweets from executives' accounts from April 21, 2013 to January 21, 2014. Our dataset comprises more than 230,000 messages sent by top executives in our sample including metadata such as user-id, time-stamp and content type. The dataset contains tweets, retweets and reply-messages from managers to other Twitter users. For career success, we collect data from the SEC filings of publicly traded companies, namely SEC Form DEF 14-A and S&P-managed ExecuComp database including salary, bonus, total value of restricted stock granted, net value of stock options exercised, and long term incentive payouts for 2013 fiscal year.

We use a supervised machine learning methodology to automatically derive the impression tactics from the sheer volume of Twitter data. Supervised learning methods require two consecutive phases. First, we compose a training set with a cost efficient and precise manual coding process. Second, we build a classification model and apply validation techniques to evaluate the performance of learning algorithm which relates the features in the primary recording to one of the class codes.

For the manual coding process, we prepare a manual coding scheme drawing on [5] and [7] to capture individual IM tactics precisely. Then we use two external coders to code a subset of tweets to form training set. We assign the tweets with an assortment of six ID numbers on the coding scheme, consisting of five IM tactics and one null ID (which is assigned when a tweet does not contain any IM tactic), with corresponding tactic explanations.

For the model building and validation process, we utilize from the training set to predict a class code for this sample. We employ machine learning and neural network classifiers in our study. Perceptron is a simple and effective type of neural network classifier, which has successfully been applied to text classification problems [18]. We observed an accuracy rate of 71 % for IM tactics classification task. Our validation process involves K-fold cross validation, which provides the overall accuracy when the model performs on an independent dataset of tweets. Next, we attempt to estimate the impact of the independent variables as five dimensions of IM tactics – ingratiation, intimidation, exemplification, supplication, and self-promotion, on career success dependent variable by using top executives as unit of analysis in a multiple regression estimation procedure.

4 Findings

We measure the effects of five IM tactics on executive career success. Our results indicate that ingratiation has a negative influence on executive career success. Although previous findings suggest that top executives are likely to engage in ingratiation strategies towards different stakeholders at different times which confer career benefits [13, 15], our viewpoint is that engaging in this strategy negatively affects the career success in the presence of context collapse (prevalent in SNP) when it is applied in front of the public eye. Next, although some researchers provided examples of how organizations use supplication strategy in social media to increase their benefits, we observed that the usage of supplication strategy is not common at the top executive level and its impact is not statistically significant. This finding indirectly supports our ex-ante position that most top executives are knowledgeable enough not to engage in supplication strategy, in order to protect their strong image in the opinion of general public.

Our empirical findings also support that self-promotion and exemplification are positively associated with career success. As we expect, we do not see the detrimental effects of context collapse on these tactics. Finally, we fail to find a statistically significant effect of intimidation on career success of a top manager. From a firm's perspective, it may be seen favorable to engage in intimidation through Twitter. However, trying to look competent, or criticizing rivals or peers over Twitter do not have a significant effect on executive career success.

By extending traditional IM theory to the online social networking context, we identify key constructs (ingratiation and supplication) that do not create a favorable image on targeted audience. The fundamental reason is the permeable boundaries among audiences. Because of the context collapse, which refers to the overlap of an individual's multiple audiences into one single platform, IM actors today need to be more careful about managing disparate impressions for their respective audiences in social networking platforms than in offline settings.

References

1. IBM: Global CEO Study (2012). ibm.com/CEOstudy
2. Bolino, M.C., Kacmar, K.M., Turnley, W.H., Gilstrap, J.B.: A multi-level review of impression management motives and behaviors. J. Manag. **34**(6), 1080–1109 (2008)
3. Goffman, E.: The Presentation of Self in Everyday Life. Anchor Books, New York (1959)
4. Wayne, S.J., Liden, R.C.: Effects of impression management on performance ratings: a longitudinal study. Acad. Manag. J. **38**(1), 232–260 (1995)
5. Bolino, M., Turnley, W.H.: Measuring impression management in organizations: a scale development based on the jones and pittman taxonomy. Organ. Res. Meth. **2**(2), 187–206 (1999)
6. Jones, E.E., Pittman, T.S.: Toward a general theory of strategic self-presentation. In: Suls, J. (ed.) Psychological Perspectives on the Self, pp. 231–262. Lawrence Erlbaum, Hillsdale (1982)
7. Mohamed, A.A., Gardner, W.L., Paolillo, J.G.: A taxonomy of organizational impression management tactics. Adv. Competitiveness Res. **7**(1), 108–130 (1999)

8. Leary, M.R., Kowalski, R.M.: Impression management: a literature review and two-component model. Psychol. Bull. **107**, 34–47 (1990)
9. Boyd, D.M.: Taken Out of Context: American Teen Sociality in Networked Publics. ProQuest (2008)
10. Marwick, A.E., Boyd, D.: I tweet honestly, i tweet passionately: twitter users, context collapse, and the imagined audience. New Media Soc. **13**(1), 114–133 (2010)
11. Westphal, J.D., Zajac, E.J.: The symbolic management of stockholders: corporate governance reforms and shareholder reactions. Adm. Sci. Q. **43**, 127–153 (1998)
12. Carter, S.M.: The interaction of top management group, stakeholder, and situational factors on certain corporate reputation management activities. J. Manag. Stud. **43**(5), 1145–1176 (2006)
13. Westphal, J.D., Stern, I.: Flattery will get you everywhere (Especially If You Are a Male Caucasian): how ingratiation, boardroom behavior, and demographic minority status affect additional board appointments at US companies. Acad. Manag. J. **50**(2), 267–288 (2007)
14. Godfrey, J., Mather, P., Ramsay, A.: Earnings and impression management in financial reports: the case of CEO changes. Abacus **39**(1), 95–123 (2003)
15. Westphal, J.D., Deephouse, D.L.: Avoiding bad press: interpersonal influence in relations between CEOS and journalists and the consequences for press reporting about firms and their leadership. Organ. Sci. **22**(4), 1061–1086 (2011)
16. Cunningham, C.: Social Networking and Impression Management: Self-presentation in the Digital Age. Rowman & Littlefield, Chicago (2013)
17. Kane, G.C., Alavi, M., Labianca, G.J., Borgatti, S.: What's different about social media networks? A framework and research agenda. MIS Q. **38**(1), 274–304 (2014)
18. Ng, H.T., Goh, W.B., Low, K.L.: Feature selection, perceptron learning, and a usability case study for text categorization. In: Proceedings of SIGIR-97, 20th ACM International Conference on Research and Development in Information Retrieval, pp. 67–73 (1997)
19. Westphal, J.D., Stern, I.: The other pathway to the boardroom: interpersonal influence behavior as a substitute for elite credentials and majority status in obtaining board appointments. Adm. Sci. Q. **51**(2), 169–204 (2006)

Does Self-promotion on Social Media Boost Career? Evidence from the Market for Executives

Yanzhen Chen[1]([⊠]), Huaxia Rui[2], and Andrew B. Whinston[1]

[1] University of Texas at Austin, Austin, TX 78712, USA
yanzhen@utexas.edu, abw@uts.cc.utexas.edu
[2] University of Rochester, Rochester, NY 14627, USA
huaxia.rui@simon.rochester.edu

Abstract. Our paper studies the impact of reputation management on executives career using the evidence from their usage of Twitter. This self-promoting behavior has both a direct influence of the barging powers in negotiating compensation and sorting in the hiring process, which helps an executive to increase the chance of finding a job. Our structural model based on a Two-Sided Matching model is able to exploit the characteristics of the other candidates to separately identify influence in barging power and sorting in getting a job. We model the assignment of executives to firms and the pay as endogenously determined. Both effects are found to be significant but in different ways. While self-promotion increases CEO compensation, in the recruiting process, only past very unsatisfactory CEO candidates benefit from reputation management. The results show more salient positive assortative matching of reputation management in Chief Marketing Officer market than the CEO market.

Keywords: Reputation management · Executive labor market · Two-sided matching · Gibbs sampling

1 Introduction

Reputation management (abbreviated to RM) help executives receive more attention, boost their perceived ability, and thus benefit their career outcome. Recent literature suggests that media visibility can improve executives job acquisition and payment premiums [5,10,11]. However, the literature has been almost silent about the mechanism how attention and perceived ability are formed. We understand attention and perceived ability as a result of both natural rewards bestowed for operating performance and more importantly in our model, strategically self-promotion. To bridge this gap in the literature, we exam whether executives are born from operating performance or made by managing reputation. Just like what makes Cinderella a princess? - Her kind heart and beauty or the gorgeous coach and crystal shoes.

© Springer International Publishing Switzerland 2016
V. Sugumaran et al. (Eds.): WEB 2015, LNBIP 258, pp. 83–96, 2016.
DOI: 10.1007/978-3-319-45408-5_8

Building on attribution theory, self-serving attributions either attribute favorable performance to internal causes (enhancing attributions) or poor performance to external causes (defensive attributions). Executives have an incentive to adjust their behavior according to such attributions to heighten (dampen) public perceptions of the association between good (bad) performance and their ability, thereby benefiting their career. Prior research finds a clear pattern that managers are more likely to attribute good news to internal causes and explain bad ones using the Dog ate my homework method [2,13,14].

Executives personal twitter accounts provide us a clean setting to capture executive self-promotion. The pioneering research on executive labor market uses executive media coverage. It can be a noisy measure since it misses the link that media coverage is influenced by executive strategic disclosure as the authors in [4] argued. Executives can strategically manage the press to encourage or avoid their name appearing in the news. To better address executive self-promotion, Blankespoor and deHaan [4] use whether an executive provides a direct quote in a firm-initiated press releases and the informativeness and vividness of the quote. Although the executives have more discretion over their comments, their speech may still largely be affected by firms group decision in firm initiated press release. However, a private twitter account is under the executives control. As an online media, Tweets reaches a board audience easily compare to the other popular social networks such as Facebook, on which information mainly spreads between friends. In addition, its 140 character limit gives the tool a different voice, a more industry based one rather than life rhapsody. All these features make Twitter a natural platform for use in self-promotion and make it a golden social media platform for self promotion for political and public figures. LinkedIn, however, is a highly structural online resume profile as a professional network, on which the content are like bullet points in a CV. Users do not have much flexibility in the content they want to promote and, therefore, is not ideal for reputation management.

The theoretical challenge in assessing the impact of RM on executive career is two-fold. First, the most commonly used methodological approach is a multi-nominal probit or logit random utility model of the firm. However, in reality, the decisions from both sides determine the employment relationship. Specifically, a companys willingness to extend an offer depends on its preference and candidate pool. Meanwhile, the willingness to join a company relies on an executive's preference and the availability of an offer. Second, both firm-executive assignments and the distribution of compensation in contracts are inherently jointly determined and thus, both are endogenous. Finding an instrument is unfortunately not straightforward since the economics underlying executive labor market makes it difficult to find a valid one. The matching of an executive with a firm is determined by their mutual decisions, and when their characteristics relate to the hiring decision and also to some extent, affect compensation, they are not valid instruments.

To overcome the problem of missing instruments, we develop a structural model that exploits the implications of compensation negotiations to separate pay determination and sorting in the recruiting process. This paper contributes to the literature by being the first to directly estimate how RM affects the executive labor market. Sorting implies an executive candidates decision depends on the characteristics of all others in the market. A given candidate is pushed down in the relative ranking and is left with worse companies if there are more capable candidates. However, the compensation is contract specific to a certain executive and thus independent of other candidates characteristics. This exogenous variation provided by other candidates help us to identify sorting and compensation negotiation, which is to the same spirit of Berry, Levinsohn, and Pakes [3]. Specifically, our structural model has two parts. The first part is the recruiting stage. It is a variation of the College Admissions Model, for which an equilibrium matching always exists [6,12]. The matching model controls for the sorting and selection of observable employment records. The second part is compensation negotiation. We jointly estimate two parts to eliminate bias due to the endogeneity problem. We obtain Bayesian inference using a Gibbs sampling algorithm [7–9] with data augmentation [1]. Estimation is numerically intensive. However, Bayesian inference is feasible using a Gibbs sampling algorithm that performs Markov chain Monte Carlo simulations.

Our estimation results provide an interesting picture about the executive market across different positions. First, reputation management impacts the two stages of job hunting differently. It increases an executive's paycheck to the same extent despite her relative performance in the last period. Whereas, in the recruiting process, only executives who did unsatisfactory jobs in the last period in the CEO market (and those who were superstars in CMO market) will gain a better chance of winning offers. Secondly, career outcomes are sensitive to RM to a different extent across positions. Self-promotion affects CMO market greater than CEO market. CMOs are more likely to attain an appreciation of perceived value by leveraging reputation management when competing with other candidates. Counterfactual experiment shows, for CMOs who did a less satisfactory job in the last period, self-promotion can help, approximately 9 % of them get a better offer according to their preferences, given other candidates remain the same.

2   Structural Empirical Model

2.1   Two-Sided Matching Model

Recruiting requires the consent of both the CEO candidates and the company. We use a two-sided matching model to capture this mutual decision and address the endogeneity problem resulting from the sorting in matching. This model is a static equilibrium model from cooperative game theory. For any firm $i \in I_m$ and $j \in J_m$, and where $\mathrm{m} = 1, 2 \ldots M$ is the market index, we have the utility function for both two sides (the executive candidates market and the hiring firms market) U_{ij}^F and U_{ij}^E. For executive applicant j, her utility U_{ij}^E of working for firm i depends on the firm characteristics F_i and her interaction with the firm N_{ij}:

$$U_{ij}^F = F_i^T \cdot \beta_1 + N_{ij}^T \cdot \beta_2 + \eta_{ij} \equiv X_{ij}^{F,T} \cdot \beta + \eta_{ij}. \tag{1}$$

where $\eta_{ij} \sim N(0, \sigma_\eta^2)$. Likewise, firm i's willingness of recruiting candidate j depends on j's characteristics and their interaction N_{ij}:

$$U_{ij}^E = E_j^T \cdot \gamma_1 + N_{ij}^T \cdot \gamma_2 + \delta_{ij} \equiv X_{ij}^{E,T} \cdot \gamma + \delta_{ij} \tag{2}$$

where $\delta_{i,j} \sim N(0, \sigma_\delta^2)$.

We capture size and profitability for firm characteristics, using ln(total asset) and average return on equity respectively. CEO characteristics consist of 5 parts: market value of the executive up to last year (Compensation in t-1 period), the performance of the executive candidate last year, RM, the interaction between performance and RM and the expected yet unearned pay from the current employer as part of opportunity cost. The performance of the executive is the change in ROE adjusted by same industry for last year. It captures the previous firm's relative market value change during the executive's working period. We use the personal Twitter account as a measure of reputation management. It is a dummy variable. To exam the different impacts reputation management has on successful (less satisfactory) executives, we include the interaction of quadratic form of executive performance and reputation management. As for the interaction between firm and executives (N_{ij}), we include geographic distance, expertise difference (whether the current employer and the potential new employ are from the same industry as defined by SIC, Standard Industrial Classification code) and attached advantage (whether she is a current employee of the firm).

Equilibrium. The authors in [14] proved in the college admissions model, a matching always exists and is pair-wise stable if and only if it is group stable. Thus, the equilibrium concept we use in our two-sided matching model is pairwise stability.

The following proposition characterizes stable equilibrium using both $\overline{U_{ij}^F}, \overline{U_{ij}^E}$ and $\underline{U_{ij}^F}, \underline{U_{ij}^E}$.

Proposition. A matching μ_m is stable if and only if the following inequality holds. We denote μ^e the equilibrium matching in market m.

$$\mu_m = \mu_m^e \Leftrightarrow U_{ij}^F < \overline{U_{ij}^F} \text{ and } U_{ij}^E < \overline{U_{ij}^E} \forall (i,j) \notin \mu_m$$
$$U_{ij}^F > \underline{U_{ij}^F} \text{ and } U_{ij}^E > \underline{U_{ij}^E}, \forall (i,j) \in \mu_m. \tag{3}$$

Specifically, for an unmatched pair (i, j), the deviation from the equilibrium is firm i is willing to hire executive j instead, and j is happy to job hopping. This can be translated to: executive j prefers firm i better than his current employer $(\mu_m(j))$: $U_{ij}^F > U_{\mu_m(j),j}^F$. At the same time firm i prefers j than the least preferred incumbent executive $(\mu_m(j))$ in the same position: $U_{ij}^E > min\left(U_{i,\mu_m(i)}^E\right)$. So (i, j) is not a blocking pair iff $U_{ij}^F < U_{\mu_m(j),j}^F$ or $U_{ij}^E < min\left(U_{i,\mu_m(i)}^E\right)$. For convenience, we define the $\bar{U}_{ij}^F$ and $\bar{U}_{ij}^E$ as follows:

$$\bar{U}_{ij}^F = \begin{cases} U_{\mu_m(j),j}^F & ,if\ U_{ij}^E > min\left(U_{i,\mu_m(i)}^E\right) \\ +\infty & otherwise \end{cases}$$

$$\bar{U}_{ij}^E = \begin{cases} min\left(U_{i,\mu_m(i)}^E\right) & ,if\ U_{ij}^F > U_{\mu_m(j),j}^F \\ +\infty & otherwise \end{cases} \tag{4}$$

So equivalently, (i, j) is not a blocking pair if and only if $U_{ij}^F < \bar{U}_{ij}^F$ and $U_{ij}^E < \bar{U}_{ij}^E$.

For a matched pair (i, j), any agent's deviation will break the equilibrium. That means:

if firm i would like to abandon its current employed executive and hire a better one j': $U_{i,j}^E < \max\limits_{j' \in f(i)} U_{i,j'}^E$, where $f(i)$ is the set of executives that are not currently work for firm i but would prefer to do so.

or executive j prefers to work for a more desirable and feasible firm i': $U_{i,j}^F < \max\limits_{i' \in f(j)} U_{i',j}^F$, where $f(j)$ is the set of firms that would rather to hire j than their current employee.

These two feasible deviation sets are given by

$$f(i) = \{j \in J_m/\mu_m(i) : U_{ij}^F > U_{\mu(j),j}^F\}$$

$$f(j) = \{i \in I_m/\mu_m(j) : U_{i,j}^E > \min\limits_{j' \in \mu_m(i)} U_{i,j'}^E\} \tag{5}$$

We define U_{ij}^F and U_{ij}^E as the follows:

$$\underline{U_{ij}^F} = \max\limits_{i' \in f(j)} U_{i',j}^F, \underline{U_{ij}^E} = \max\limits_{j' \in f(i)} U_{i,j'}^E \tag{6}$$

Therefore, either executive j or firm i will block the pair if and only if $U_{ij}^F > \underline{U_{ij}^F}$ and $U_{ij}^E > \underline{U_{ij}^E}$.

2.2 Compensation

Once firm i and executive candidate j express mutual interest to form a pair, the two players are going to negotiate the compensation based on the characteristics of both sides and their interactions. We characterize the compensation negotiation process as:

$$r_{ij} = \alpha_0 + F_i^T \cdot \alpha_1 + E_j^T \cdot \alpha_2 + N_{ij}^T \cdot \alpha_3 + \epsilon_{ij} \equiv W_{ij}^T \cdot \alpha + \epsilon_{ij}. \tag{7}$$

where $\epsilon_{ij} \sim N(0, \sigma_\epsilon^2)$, r_{ij} is the compensation we observe, all other covariates are defined as in (1).

As we have discussed in the introduction, the error term in the compensation negotiation process is correlated with the ones in stage 1. We decomposed the error term in the compensation negotiation process into orthogonal terms as

$$\epsilon_{ij} = k \cdot \eta_{ij} + \lambda \cdot \delta_{ij} + v_{ij}. \tag{8}$$

Since theory predicts that executives with higher unobserved qualities have stronger bargaining power in compensation negotiation. So λ is positive.

We set σ_δ^2 and σ_η^2 to be 1 to fix the scales and exclude constant terms to fix the levels. Thus the joint distribution of η_{ij}, δ_{ij} and v_{ij} is

$$\begin{pmatrix} \epsilon_{ij} \\ \eta_j \\ \delta_i \end{pmatrix} \sim \begin{pmatrix} \kappa^2 + \lambda^2 + \sigma_v^2 & \kappa & \lambda \\ \kappa & 1 & 0 \\ \lambda & 0 & 1 \end{pmatrix} \tag{9}$$

This is without loss of generality, as the joint distribution of ϵ_{ij}, η_{ij} and δ_{ij} remains unchanged.

Based on the data generation process, the augmented posterior density is

$$p(U^F, U^E, \gamma, \beta, \theta | W, r, \mu) =$$
$$\phi_0(\theta) \cdot \prod_m \{[\prod_{(i,j) \in \mu_m} \phi((r_{ij} - W_{ij}^T \cdot \alpha - \kappa \cdot (U_{ij}^F - X_{ij}^{F,T} \cdot \beta)$$
$$- \lambda(U_{ij}^E - X_{ij}^{E,T} \cdot \gamma))/\sigma_v) \cdot 1(U_{ij}^F > U_{ij}^F) \cdot \phi(U_{ij}^F - X_{ij}^{F,T} \cdot \beta)$$
$$\cdot 1(U_{ij}^E > U_{ij}^E) \cdot \phi(U_{ij}^E - X_{ij}^{E,T} \cdot \gamma)]$$
$$\cdot [\prod_{(i,j) \notin \mu_m} 1(U_{ij}^F < \bar{U}_{ij}^F) \cdot \phi(U_{ij}^F - X_{ij}^{F,T} \cdot \beta) \cdot 1(U_{ij}^E < \bar{U}_{ij}^E) \ \phi(U_{ij}^E - X_{ij}^{E,T} \cdot \gamma)]$$
$$\tag{10}$$

Here θ stands for the parameters we need to estimate: $(\alpha, \beta, \gamma, \kappa, \lambda, \sigma_v^2)$. μ_m is the observed firm-executive matching pair in market m. U^E, U^F, W, r, μ contain all U_{ij}^E, U_{ij}^F, W_{ij}, r_{ij}, μ_m in all markets. $1(.)$ is the indicator function.

2.3　Structural Identification

The theoretical challenge of our research question arises from sorting and interaction between agents in both sides of the markets. When an executive signs a contract with a firm, it will greatly reduce the probability that other candidates can join this company, due to the quota of each company. So it is improper to analyze each candidates decision independently. As a result, the likelihood cannot factor into a product over the individual choice likelihood. Rather, this fundamental property of our research question requires all error terms to be integrated simultaneously. The maximum likelihood function will suffer from the curse of dimensionality and is impossible for estimation. However, Bayesian estimation using Gibbs sampling and data augmentation provides a solution to this problem.

The prior distributions of $(\alpha, \beta, \gamma, \kappa, \lambda, \sigma_v^2)$ are denoted $N(\mu_\alpha, \Sigma_\alpha)$, $N(\mu_\beta, \Sigma_\beta)$, $N(\mu_\gamma, \Sigma_\gamma)$, $N(\mu_\kappa, \Sigma_\kappa)$ and $\sigma_v^2 \sim InvG(a, b)$ respectively. As theory predicts, γ stands for the positive correlation between unobserved executive quality and her bargaining power. We set the prior for γ as a truncated normal distribution, $\lambda \sim TN(0, \sigma_\lambda^2, +)$.

In the estimation procedure, the prior distributions have means of zero and variances of 100. For all estimated parameters, the prior variances are more than 100 times greater than the posterior variances. This suggests the Gibbs sampling is well behaved and the posterior distribution reflects the information from data well.

2.4 Conditional Distributions of Latent Scores/Utility

The conditional augmented posterior distribution of U_{ij}^F and U_{ij}^E varies according to whether firm i and executive candidate j are matched or not. When $(i, j) \notin \mu_m$, the densities are simply:

$$p(U_{ij}^F | X_{ij}^F, U_{-ij}^F, U_i^E, \beta) = 1(U_{ij}^F < \bar{U}_{ij}^F) \cdot \phi(U_{ij}^F - X_{ij}^{F,T} \cdot \beta) \tag{11}$$

$$p(U_{ij}^E | X_{ij}^E, U_{-ij}^E, U^F, \gamma) = 1(U_{ij}^E < \bar{U}_{ij}^E) \cdot \phi(U_{ij}^E - X_{ij}^{E,T} \cdot \gamma) \tag{12}$$

When $(i, j) \in \mu_m$, the real compensation of matching pair is observed. The correlation between the error terms provide additional information about the latent scores, and the conditional densities are

$$\begin{aligned} &p(U_{ij}^F | X_{ij}^E, X_{ij}^F, U_{-ij}^F, U_{ij}^E, \gamma, \beta, r_{ij}, W_{ij}, \alpha, \lambda, \kappa, \sigma_v^2) \\ &= 1(U_{ij}^F > \underline{U}_{ij}^F) \cdot N(U_{ij}^F; \mu_{U_{ij}^F}, \sigma_{U_{ij}^F}^2) \end{aligned} \tag{13}$$

$$\begin{aligned} &p(U_{ji}^E | X_{ij}^E, X_{ij}^F, U_{ij}^F, Q_{-ji}^E, \gamma, \beta, r_{ij}, W_{ij}, \alpha, \lambda, \kappa, \sigma_v^2) \\ &= 1(U_{ij}^E > \underline{U}_{ij}^E) \cdot N(U_{ij}^E; \mu_{Q_{ji}^E}, \sigma_{U_{ji}^E}^2) \end{aligned} \tag{14}$$

where

$$\begin{aligned} \mu_{U_{ij}^E} &= \big[\lambda \cdot (r_{ij} - W_{ij}^T \cdot \alpha - \kappa \cdot (U_{ij}^F - X_{ij}^{F,T} \cdot \beta) \\ &\quad + \lambda \cdot X_{ij}^{E,T} \cdot \gamma)/\sigma_v^2 + X_{ij}^{E,T} \cdot \gamma\big]/\left(\lambda^2/\sigma_v^2 + 1\right) \\ \sigma_{U_{ij}^E}^2 &= \frac{\sigma_v^2}{\sigma_v^2 + \lambda^2} \\ \mu_{U_{ji}^F} &= \big[\kappa \cdot (r_{ij} - W_{ij}^T \cdot \alpha + \kappa \cdot X_{ij}^{F,T} \cdot \beta_j \\ &\quad - \lambda \cdot (U_{ij}^E - X_{ij}^{E,T} \cdot \gamma))/\sigma_v^2 + X_{ij}^{F,T} \cdot \beta\big]/\left(\kappa^2/\sigma_v^2 + 1\right) \\ \sigma_{U_{ij}^F}^2 &= \frac{\sigma_v^2}{\sigma_v^2 + \kappa^2} \end{aligned}$$

These are the normal distributions that truncated from above (below). The expressions for $\bar{U}_{ij}^F, \bar{U}_{ij}^E, \underline{U}_{ij}^F$ and $\underline{U}_{ij}^E$ are given in Eqs. 4 and 6.

The information we use to update the conditional augmented posterior distribution of preference coefficients of the two sides: β and γ comes from two sources: The latent score/preference of the firm (executive) side of the market and the observed compensation of those matching pairs. So the density of the conditional distributions are

$$p(\beta|.) \sim N(\hat{\mu}_\beta, \hat{\Sigma}_\beta) \tag{15}$$

$$p(\gamma|.) \sim N(\hat{\mu}_\gamma, \hat{\Sigma}_\gamma) \tag{16}$$

where

$$\hat{\mu}_\beta = \hat{\Sigma}_\beta \cdot [\Sigma_{\beta,}^{-1} \cdot \mu_\beta + \sum_{m=1}^{M} (\sum_{(i,j)\in\mu_m}(\frac{-1}{\sigma_v^2} \cdot \kappa \cdot X_{ij}^F \cdot (r_{ij} - W_{ij}^T \cdot \alpha - \kappa \cdot U_{ij}^F - \lambda(U_{ij}^E$$

$$-X_{i,j}^{E,T} \cdot \gamma)) + X_{ij}^F \cdot U_{ij}^F) + \sum_{(i,j)\notin\mu_m} X_{ij}^F \cdot U_{ij}^F)]$$

$$\hat{\Sigma}_\beta = [\Sigma_\beta^{-1} + \sum_{m=1}^{M} (\sum_{(i,j)\in\mu_m} \frac{1}{\sigma_v^2} \cdot (\sigma_v^2 + \kappa^2) \cdot X_{ij}^F \cdot X_{ij}^{F,T} + \sum_{(i,j)\notin\mu_m} X_{ij}^F \cdot X_{ij}^{F,T})]^{-1}$$

$$\hat{\mu}_\gamma = \hat{\Sigma}_\gamma \cdot [\Sigma_\gamma^{-1} \cdot \mu_\gamma - \sum_{m=1}^{M} (\sum_{(i,j)\in\mu_m}(\frac{-1}{\sigma_v^2} \cdot \lambda \cdot X_{ij}^E \cdot (r_{ij} - W_{ij}^T \cdot \alpha$$

$$-\kappa \cdot (U_{ij}^F - X_{ij}^{F,T} \cdot \beta) - \lambda U_{ij}^E)) - X_{ij}^E \cdot U_{ij}^E) - \sum_{(i,j)\notin\mu_m} X_{ij}^E \cdot U_{ij}^E)]$$

$$\hat{\Sigma}_\gamma = [\Sigma_\gamma^{-1} + \sum_{m=1}^{M} (\sum_{(i,j)\in\mu_m} \frac{1}{\sigma_v^2} \cdot (\sigma_v^2 + \lambda^2) \cdot X_{ij}^E \cdot X_{ij}^{E,T} + \sum_{(i,j)\notin\mu_m} X_{ij}^E \cdot X_{ij}^{E,T})]^{-1}$$

The conditional posterior distributions of the coefficients in compensation determination stage depends on the prior and information from the observed compensation. Collecting terms involving α in (1.6) and complete squares, this distribution is as follows:

$$\alpha \sim N(\hat{\mu}_\alpha, \Sigma_\alpha) \tag{17}$$

where

$$\hat{\mu}_\alpha = \hat{\Sigma}_\alpha \cdot \left(- (\Sigma_\alpha)^{-1} \cdot \mu_\alpha + \frac{1}{\sigma_v^2} \sum_{m=1}^{M} \sum_{(i,j)\in\mu_m} (r_{ij} - \kappa \cdot (U_{ij}^F \right.$$

$$\left. -X_{ij}^{F,T} \cdot \beta) - \lambda(U_{ij}^E - X_{ij}^{E,T} \cdot \gamma)) \cdot W_{ij} \right)$$

$$\hat{\Sigma}_\alpha = \left((\Sigma_\alpha)^{-1} + \frac{1}{\sigma_v^2} \sum_{m=1}^{M} \sum_{(i,j)\in\mu_m} W_{ij} \cdot W_{ij}^T \right)^{-1}$$

Update the inference of κ and λ requires the information from the observed compensation. The densities of these conditional distribution are normal distributions, of which λ is truncated from below at zero. It is because we assume any unobserved characteristics that makes the candidate more appealing in the eyes of a certain firm will increase the executive's bargaining power with this firm. The conditional posterior distributions are:

$$\kappa \sim N(\hat{\mu}_k, \hat{\sigma}_k), \quad \lambda \sim TN(\hat{\mu}_\eta, \hat{\sigma}_\eta, +) \tag{18}$$

where

$$\hat{\mu}_k = \hat{\sigma}_k^2 \cdot \sum_{m=1}^{M} \sum_{(i,j)\in\mu_m} \frac{(U_{ij}^F - X_{ij}^{F,T}\cdot\beta)\cdot(r_{ij} - W_{ij}^T\cdot\alpha - \lambda(U_{ij}^E - X_{ij}^{E,T}\cdot\gamma))}{\sigma_v^2}$$

$$\hat{\sigma}_\kappa^2 = \left[\sum_{m=1}^{M} \sum_{(i,j)\in\mu_m} \left(\frac{U_{ij}^F - X_{ij}^{F,T}\cdot\beta}{\sigma_V} \right)^2 + \frac{1}{\sigma_k^2} \right]^{-1}$$

$$\hat{\mu}_\lambda = \hat{\sigma}_\lambda^2 \cdot \sum_{m=1}^{M} \sum_{(i,j)\in\mu_m} \frac{(U_{ij}^E - X_{ij}^{E,T}\cdot\gamma)\cdot(r_{ij} - W_{ij}^T\cdot\alpha - \kappa(U_{ij}^E - X_{ij}^{E,T}\cdot\gamma))}{\sigma_v^2}$$

$$\hat{\sigma}_\lambda^2 = \left[\sum_{m=1}^{M} \sum_{(i,j)\in\mu_m} \left(\frac{U_{ij}^E - X_{ij}^{E,T}\cdot\gamma}{\sigma_v} \right)^2 + \frac{1}{\sigma_\lambda^2} \right]^{-1}$$

The conditional posterior distribution of σ_v^2 is $InvG(\hat{a}_{\sigma_v^2}, \hat{b}_{\sigma_v^2})$, updated using the observed compensation.

Where

$$\hat{a}_{\sigma_v^2} = a_{\sigma_v^2} + \frac{\sum_{m=1}^{M} |\mu_m|}{2}$$

$$\hat{b}_{\sigma_v^2} = b_{\sigma_v^2} + \sum_{m=1}^{M} \sum_{(i,j)\in\mu_m} \left(r_{ij} - W_{ij}^T\cdot\alpha - \kappa\cdot(U_{ij}^F - X_{ij}^{F,T}\cdot\beta) \right.$$
$$\left. -\lambda(U_{ij}^E - X_{ij}^{E,T}\cdot\gamma) \right)^2 / 2$$
$$|\mu_m| : number\ of\ elements\ in\ the\ set\ \mu_m$$

3   Data

We study data on executive employment records and compensation of S&P 500 constituents companies from 2010 to 2013. Especially, we exam the impact of reputation management on CMO (marketing and sales related chief officers) and CEO markets. We exclude firms that operate in certain industrial sectors: financial corporations (SIC(6000-6999)) and regulated utilities (4900-4999). Since profitability and valuation data for financial firms are difficult to calculate and to compare with firms in other sectors. Similar concerns apply for regulated utilities. Their profitability and valuation can be strongly influenced by government regulations. The executives personal twitter accounts and creation dates are hand collected and double verified by two independent research assistants. We excluded firm twitter accounts and accounts created by the executives fans under her name by twitter accounts verification feature and going through their tweets. We treat each employment record as an observation. For example, our final sample of CMO market consists of 592 matching pairs and compensation contracts from 290 different executives. We divide markets using positions and years. For job changings happen in between a year we consider the later matching with the compensation amount equal to the sum of the two paychecks. Since there can be unusual compensation changes (for example, payments in event of involuntary termination of employment) or unobserved switching cost/gain for the executive in the paycheck. Thus, a weighted sum of compensation using working duration will introduce bias. On the other hand, when negotiating the pay with the next employer, both sides will take the changes in previous compensation into consideration. Thus, the sum of the two payments will be a reasonable estimation of the amount she can get if she started to work in the new company for a whole year. Our other financial data is from Compustat database.

4   Results

4.1   Empirical Findings

Using a sample of CEOs that worked for S&P 500 companies between 2010 and 2013, our joint estimation results are reported in Tables 1 and 2. The results are based on 50,000 draws from which the initial 25,000 are burn-in. Visual inspection of the draws shows that convergence to the stationary posterior distribution occurs for most parameters. We plot 4 of them from both stages of the model in Fig. 1. We standardized dependent variable and all non-dummy covariates before analysis.

Table 1. Estimates of utility equations

CEO utility			Firm utility		
	MEAN	95 % HDPI		MEAN	95 % HDPI
Ln(Asset)	1.230 **	(0.720, 1.651)	Unearn	0.687	(−0.785, 1.640)
ROE	0.3180	(−0.457, 0.967)	Pay t−1	3.921**	(2.112, 6.125)
Dist	−0.395**	(−0.636, −0.129)	Rept Mgmt(RM)	−0.280	(−1.249, 0.556)
Experience	5.291**	(4.125, 6.688)	Exec Perfor(EP)	0.570**	(0.382, 1.274)
Attached	−1.223**	(−1.850, −0.397)	RM × EP2 (+)	1.022	(−0.492, 2.349)
			RM × EP2 (−)	2.684**	(1.2780, 4.725)
			Dist	−0.386**	(−0.679, −0.093)
			Experience	4.900**	(2.549, 6.902)
			Attached	0.0572	(−1.035, 1.204)

Table 2 shows that reputation management is significantly positive at the 5 % level. This result reveals self-promoting boosts CEO compensation. The interactions of RM and quadratic term of past performance-whether the CEO outperformed the past relative to the industry in regard to returns for shareholders, are not significant. It shows, conditional on getting an offer, RM equally benefits CEO's compensation. Differently, in Table 1 we can see only the interaction involves executive performance when the executive under-performed the past year is positive significant. This shows in recruiting process, RM only helps those less satisfactory executives to defense themselves and make them more attractive, but will not be icing on the cake for those successful ones. The other control variables in executive utility equation show executive candidates prefer

Table 2. Estimates of compensation equations

Compensation					
	MEAN	95 % HDPI		MEAN	95 % HDPI
Const	0.043	(−0.454, 0.325)	Exec Perfor(EP)	0.312**	(0.177, 0.693)
Ln(Asset)	0.080**	(0.030, 0.129)	RM EP2 (+)	−0.0707	(−2.530, 3.012)
ROE	0.010	(−0.083, 0.101)	RM EP2 (−)	0.015	(−0.050, 0.081)
Unearn	0.321**	(0.087, 0.545)	Dist	0.0303	(−0.004, 0.064)
Pay t−1	0.417**	(0.319, 0.514)	Experience	−0.163	(−0.997, 1.454)
Rept− Mgmt(RM)	0.136**	(0.001, 0.271)	Attached	0.216	(−0.774, 1.303)
κ	0.041**	(4.937E−07, 0.091)			
λ	0.059**	(1.053E−05, 0.161)			
σ_v^2	0.112	(0.087, 0.136)			

large companies as the ln(asset) is positively significant. They also assertively match to firm that are geographically closer and is in the same industry as their current employer. The firm utility equation estimations reveal that firms favor candidates that have high market value and outperformed the past when worked with current employer. They also favor candidates that have experience in the same industry and hesitate in hiring someone from a long distance. Comparison of variable attach in both equations shows that executives are more willing to change an employer than the firms do. Control variables in compensation equation show that big firms can afford to pay more and executive who have higher market value in the past tend to get higher compensation in the future and are significantly positive provides evidence that hiring process and compensation stages are correlated with each other. This proves reduced form regression on firm, executive attributes as well as their interactions will suffer from the endogenous problem. It further strengthens that our model addresses the research question well. Relatively small shows that our model fits the data well. We also fit the model using the data from CMO markets from 2010 to 2013. The results basically remain unchanged except for in the firm utility equation, both interaction terms when the executive performed well or not are positively significant (2.3984 and 4.1009 respectively). This suggests CMO market is more sensitive to reputation management. Both types of executives can improve their situations when competing for an offer and it benefit less successful executives more than the other type.

4.2 Counterfactual Experiment

Using the estimates of the model, we perform counterfactual analysis to investigate how reputation management benefits executives in job hunting. Estimation from both markets consistent with each other in that RM helps candidates who fail to improve the firms relative performance last period to be more competitive in winning an offer. Therefore, we evaluate this counterfactual career outcome using those less successful candidates that do not have a twitter account from

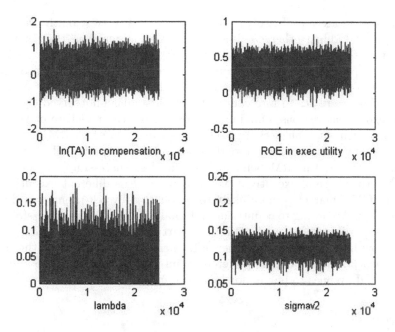

Fig. 1. Convergence.

CMO market. Specifically, we use the estimated parameters in firm utility equation to evaluate the counterfactual utility of hiring an executive who was less successful in the last period as if they were in the self-promotion business last period. We then use the counterfactual utilities of candidate j with every potential firm in the market to re-run the matching process, holding all else unchanged. As a result, we can get a counterfactually stable matching result of the given market. In this way, we can compare the counterfactual matching firms with the real matching results to see if the counterfactual employer ranks higher according to the executives preference. We perform the above experiment for CMO candidates who did a less satisfactory job in the last period and did not promote themselves. Results show self-promotion can help 9.15 % of them get better offers according to their preferences, given other candidates remain the same.

5   Conclusions

Perceived reputation and ability are a vital factor in deciding an executives career outcomes. However, prior literature has not yet examined how such perception forms. We model an executive as both born from achieved operating performance, and further, at least partly, made by reputation management. We introduce a better measure of RM: whether a candidate involves in twitter broadcasting. This is a better proxy compared to the vanguard works using media coverage and firm initiated press since a personal account mostly reflects the executives viewpoints and explanations rather than the firms or other groups. Our model is able to separately evaluate impacts of RM on the two jointly determined outcomes: outbidding other candidates and adding zeros on their paychecks. We leverage the interaction in decision making between executives to solve the endogeneity problem. That is, because of sorting in the recruiting process, the existence of other candidates affects the matching results and leads firms with different characteristics to hire candidates with similar unobserved characteristics for exogenous reasons. Our model provides a clearer picture of how reputation management influences an executives career. We find RM increases an executives pay regardless how he performed in the past period. However, in the recruiting process, this RM behavior benefits less successful candidates more than the other type of executives. We also find across different executive positions, the CMO market is more likely to be affected by self-promotion than the CEO market. According to counterfactual analysis, 9.15 % less successful CMOs could have achieved a better position by starting self-promotion. Although our model provides a very useful framework, there are limitations to the current model. Incorporating dynamic preferences into the model would be one of the other fruitful avenues to pursue for future research.

References

1. Albert, J., Chib, S.: Bayesian analysis of binary and polychotomous response data. J. Am. Stat. Assoc. **88**, 669–679 (1993)
2. Baginski, S., Hassell, J., Kimbrough, M.: Why do managers explain their earnings forecasts? J. Account. Res. **42**(1), 1–29 (2004)
3. Berry, S., Levinsohn, J., Pakes, A.: Automobile prices in market equilibrium. Econometrica **63**, 841–890 (1995)
4. Blankespoor, E., deHaan, Ed.: CEO Visibility: Are Media Stars Born or Made? Rock Center for Corporate Governance at Stanford University Working Paper No. 204, Stanford University Graduate School of Business Research Paper No. 15–21 (2015)
5. Falato, A., Li, D., Milbourn, T.: Which Skills Matter in the Market for CEOs? Evidence from Pay for CEO Credentials. Working Paper (2014)
6. Gale, D., Shapley, L.: College admissions and the stability of marriage. Am. Math. Monthly **69**, 9–15 (1962)
7. Gelfand, A., Smith, A.: Sampling-based approaches to calculating marginal densities. J. Am. Stat. Assoc. **85**, 398–409 (1990)
8. Geweke, J.: Using simulation methods for bayesian econometric models: inference, development, and communication. Econometric Rev. **18**(1), 1–73 (1999). (with discussion and rejoinder)
9. Geweke, J., Gowrisankaran, G., Town, R.: Bayesian inference for hospital quality in a selection model. Econometrica **71**(4), 1215–1239 (2003)
10. Malmendier, U., Tate, G.: Superstar CEOs. Q. J. Econ. **124**, 1593–1638 (2009)
11. Rajgopal, S., Shevlin, T., Zamora, V.: CEOs' outside employment opportunities and the lack of relative performance evaluation in compensation contracts. J. Finance **61**, 1813–1844 (2006)
12. Roth, A., Sotomayor, M.: Two-Sided Matching: A Study in Game-Theoretic Modeling and Analysis. Econometric Society Monograph Series. Cambridge University Press, Cambridge (1990)

13. Salancik, G., Meindl, J.: Corporate attributions as strategic illusions of management control. Adm. Sci. Q. **29**(2), 238–254 (1984)
14. Staw, B., McKechnie, P., Puffer, S.: The justification of organizational performance. Adm. Sci. Q. **28**(4), 582–600 (1983)

The Ethics of Online Social Network Forensics

Jongwoo Kim[1(✉)], Richard Baskerville[2], and Yi Ding[3]

[1] Department of MSIS, University of Massachusetts Boston, Boston, MA, USA
jonathan.kim@umb.edu
[2] Department of CIS, Georgia State University, Atlanta, GA, USA
baskerville@gsu.edu
[3] School of Science and Technology, Georgia Gwinnett College, Atlanta, GA, USA
yding1@ggc.edu

Abstract. Online social networks (OLSNs) are electronically-based social milieux where individuals gather virtually to socialize. The behavior and characteristics of these networks can provide evidence relevant for detecting and prosecuting policy violations, crimes, terrorist activities, subversive political movements, etc. Some existing forensics methods and tools are useful for such investigations. Further, forensics researchers are developing new methods and tools specifically for investigating and examining online social networks. While the privacy rights of individuals are widely respected, the privacy rights of social groups are less well developed. In the current development of OLSNs and computer forensics, the compromise of group privacy may lead to the violation of individual privacy. This paper examines the ethics regarding forensics examinations of online social networks. As with individual privacy, ethical tensions exist in social group privacy between the privacy rights that can be afforded to the membership, and the rights of institutions to detect and govern conspiracies to subversion, crimes, discrimination, etc.

Keywords: Ethics · Privacy · Online social network · Computer forensics

1 Introduction

There has been considerable recent expansion in the development of OLSNs. This development is paralleled by improving investigative and analytic techniques for dealing with large bodies of unstructured data [1]. In particular the digital forensics field has developed tools that are applicable to forensics analysis of such social networks [2–4]. Such digital forensics analysis has traditionally been conducted within police frameworks. However, digital forensics analysis is increasingly being used for internal organizational investigations and incident response. An example of the latter is the use of digital forensics for the purpose of electronic discovery under circumstances of civil lawsuits (such as wrongful dismissal). Many countries have well-defined laws dealing with legal rights to conduct such investigations. Similarly, there are established international privacy rights frameworks that protect personal data about individuals. While it is clear that individuals have certain rights to the proper governance of their personal data, to what degree do these privacy rights extend to social groups?

© Springer International Publishing Switzerland 2016
V. Sugumaran et al. (Eds.): WEB 2015, LNBIP 258, pp. 97–111, 2016.
DOI: 10.1007/978-3-319-45408-5_9

Most OLSNs must ultimately have owners that sponsor the social network's servers for commercial or social purposes. In administering access, these owners may collect minimal personal data about the users (name, email, address, etc.). As a consequence, the owner's data files may not hold much risk for the privacy of the individuals or the social group itself (except perhaps for the harvesting of email addresses for spamming). However, there is much higher risk in the various posts made by the individuals to their profiles or as communication for other members of the social group.

The privacy and ethics issues engulfing online social networks are complex. Basic social notions are shifting as a result of the technology use. For example, there are concerns that OLSNs are changing the ethics of friendship [5]. Superficial online friendships may fail to meet the shared life criterion of virtue friendship [6]. But this definition of friendship goes beyond philosophy. Friendship is used to provide the conceptual integrity in someone's online presence that often distinguishes who is let into one's privacy and who is excluded [7].

The release of profile data about communities online has encountered early difficulties with anonymizing such data. For example, the release of Facebook profile data was quickly mined to reveal the underlying community. There were concerns that, although the profiles did not contain any obvious personally identifying data, the profiles could be pierced using features such as physical, physiological, mental, economic, cultural or social identity. Such piercings could enable re-identification of individuals [8]. The presence of such features in online social networks further complicates the need to govern privacy within national identity systems [9]. There appears to be a tendency for online social networks to pressure participants toward self-disclosure of private information [10]. Governments may be required in future to regulate social networks because of the ease by which participants may victimize themselves through unnecessary self-disclosures [11].

Like friendship, the very shape of individual identity may evolve as a result of OLSNs. Another ethical dimension is that of identity and the potential for online social networks to either reinforce or deteriorate the identity of the participant [12]. Beyond actualizing social concepts such as friendship, and privacy/identity disclosures, OLSN participation may actually redefine what these concepts mean and the norms that revolve around them.

Businesses are moving to take advantage of the data generated by social network groups. Community relations efforts strategize the promotion of products and services through social networks. Social network group data have also been used for understanding the voters' preferences in political campaigns. It will be no surprise that computer forensics can retrieve data from OLSNs. Data scientists with business intelligence capabilities can use such data to analyze and understand group characteristics (e.g., culture, movement, interest, etc.) for such purposes as social media marketing [13]. The question is whether these attributes are something that the group as a whole wants to reserve as private. Without appropriate protection of the group data, the group values might be vulnerable to misuse or subject to various threats and attacks from unexpected sources that gain access to these data and have the capability for their abuse. Furthermore the violation of group privacy may lead to the compromise of individual privacy.

The purpose of this paper is to provide a process framework that explains how network forensics is changing the vulnerability of privacy in online social networks. The interrogative framework shown at Fig. 1 provides an agenda for new directions of research in the areas of the ethics of forensics and online privacy. The arrows represent the progression of research questions that proceed from the investigation and analysis of OLSNs by computer forensics professionals. During this process, data on group and group members may be revealed thereby raising group/individual privacy issue. Our research questions originated from the framework are as follows.

1. How do commercial motives work as a background force in OLSNs privacy?
2. How can computer forensics be applied to OLSNs?
3. What privacy issues arise when OLSNs' data is acquired and analyzed?
4. Do OLSNs have group privacy?
5. What is the relationship between group privacy and individual privacy under the context of OLSNs and their privacy violation?

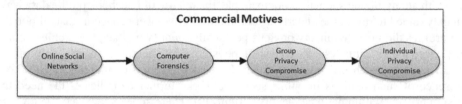

Fig. 1. Interrogative framework of OLSN privacy

We discuss our research questions in the following sections. We first start with commercial motives related to the compromise of OLSN privacy and an overview of online social networking. Then, we review computer forensics technologies and their applications to online social networks. Afterwards we discuss legal and ethical aspects of these privacy issues. Finally, we conclude our discussion.

2   Commercial Motives for OLSN Privacy Compromise

There may be strong commercial interest in minimizing group privacy for OLSNs. Businesses recognize OLSNs as effective marketing tools. Organizations are increasingly interacting with customers via social network sites [14].

Social network groups play a vital role in changing consumer behavior. Consumers today connect with brands in fundamentally different ways [15]. Prior to Internet age, consumers started with many potential brands in mind and narrow down their choices until they decided one to buy. Recent research indicates that consumers add and subtract brands from a group under consideration during an extended evaluation phase [16]. They rely heavily on digital interactions before they actually purchase. Even after a purchase, they still evaluate a shifting array of options and stay engaged with the brand via online media. Social media has been making the evaluate and advocate stages increasingly relevant.

But such sophisticated network forensics when coupled with big data analytics, data science, and data mining systems enable organizations to profile social network groups and raise digital privacy issues. Even in terms of individual privacy, few firms have developed ethical frameworks and privacy protection strategies based on multiple factors such as national history, culture, legislation, firm-specific orientation, and nature of data [17]. Organizations may find little incentives to make privacy a priority unless they are thrust into the public limelight in newspaper articles [18]. While there is a degree of business research about individual privacy [19], there is less work still about group privacy. In the networked world, organizations may need to be more proactive in responding to privacy issues related to OLSN and anticipating the problems arising from the business use of network forensics technology.

3   Online Social Networks (OLSNs)

OLSNs are online service platforms that facilitates the building of social relations among people who want to share their interests, activities, and real-life connections [20]. OLSNs consist of members' profiles, their social links, and a list of additional services that allow users to share information within their networks. Efforts to support computer-mediated social interaction and networking began in late 1970s. Early OLSNs started as generalized online communities to enable people to interact with each other through chat rooms. With new technological features such as user profile creation and friend management, a new wave of OLSNs has gained popularity since early 2000s.

Most OLSNs allow users to use three core features: (1) construct a public or semi-public profile within a bounded system, (2) articulate a list of other users with whom they share a connection, and (3) view and traverse their list of connections and those made by others within the system. Beyond sharing profiles, OLSNs vary greatly in their features - meeting new friends or dates (e.g. Friendster, Orkut), finding new jobs (e.g. LinkedIn), receiving/providing recommendations (e.g. Tribe), etc. Some note that OLSNs and online community services are different in that OLSNs are individual-centered whereas online community services are group-centered [21]. However, online community services are considered as OLSNs in a broad sense.

OLSNs are by no means business-free spaces. OLSNs have fundamentally changed the playing field for community and public relations. These online communities can intensify or destroy an organization's community relations; and to do so with astonishing speed [22]. Organizations are quickly learning that their engagement with OLSNs must not just be active, but intensive.

Prior research focusing on OLSN trust and privacy was built upon social network theory. Particularly studied have been the role of trust and intimacy, safety of young users, and representing and harvesting OLSN profile information [23]. Social network theory [24] has been applied to analyze the relevance of relations of different depth and strength under this context. Social network theory regards social relationships as a network consisting of nodes and ties. Nodes correspond to individuals within networks and ties correspond to relationships between individuals. Social networks operate at different levels (e.g. individual, groups, organizations). Social network theory can also be used to develop a consistent method in determining whether an individual has a reasonable expectation of privacy in legal cases [25]. The theory can be used to determine whether information is considered private or public. Although social network theory explains how privacy is not only an individual-level issue but also a group-level issue, prior research has focused on individual-level privacy.

4 Network Forensics

Computer forensics technology is developing impressive capabilities for revealing analytical evidence through online investigations. According to the Merriam-Webster Online Dictionary, the term "forensics" means "the application of scientific knowledge to legal problems; especially: scientific analysis of physical evidence (as from a crime scene)" [26]. The use of computer technologies has often left digital information that can be subject to forensics analysis.

Because computer devices are increasingly networked, "network-based attacks targeting critical infrastructure such as power, health, communications, financial, and emergency response services" has become a major concern [27, p. 9]. Gathering network data in a "forensically sound" process concomitantly becomes important [28, p. 1]. Compared with collecting the evidence on a local hard drive, collecting the evidence on a network is more challenging as the network data are typically "dynamic and volatile," and often require "specialized knowledge of tools and the underlying network technology" [27, p. 383].

Log analysis is a classic approach in network forensics analysis. This is because logs such as firewall logs or system event logs often consist of information about connections, user logon records, system events, and dates and times that have important evidential value in network investigation [29]. Such analysis can be performed to understand network traffic patterns, detect potential network intrusions, etc. Investigators use a variety of tools such as network traffic recorders (e.g. tcpdump, ethereal, etc.) [30], system registry data editors (e.g. RegMon, Process Explorer, etc.) [31], system event analyzers (e.g. EventLog Analyzer, etc.) [32], etc.

Network forensics analysis is still in early stages of development. However, it appears that this developing technology could provide a means for considerable investigative capabilities into OLSNs.

5 Forensic Analysis of Online Social Networks

While the ethical correctness of OLSN forensic investigation is debatable, the capability to intrude on these networks using forensics is clear. To forensics analysts, information posted on social networking sites can be valuable for such purposes as tracking policy violations, criminal groups, terrorists, political movements, etc. However, because the intense interest in OLSNs is so new, little research has been conducted so far that leads to a good understanding of the relationship between data forensics and social networks. Because forensics methods and technologies appear potentially capable for analyzing these social networking data, we briefly review these methods and technologies in the context of OLSN usage.

The forensics field is already at work developing methods and tools for forensics analysis of OLSNs. In one study, data mining methods were applied to extract the social network structure based on the online information posted on social networking sites. The focus of the analysis is on the interaction and dynamics of the network [33]. Examples of how this forensics analysis can focus include:

- Centrality in a social network
- Identification of levels of culpability for action(s) involving a group, e.g. dissemination of malicious emails
- Commitment in exchange relationships, e.g. how and when individuals become committed to a group (like pedophiles exchanging images)
- Roles of individuals in social network, e.g. Bridges, power relationships, strength of weak ties, etc.

The outliers within these social groups might also provide interesting information to forensic analysts as these outliers can indicate suspicious patterns [34].

Forensics technology makes it feasible to examine an online social group to discover its collective profile, perhaps indirectly. For example, the collective entertainment preferences might be inferred from frequent posts of film names, book titles, etc. Collective political preferences might be inferred from frequent positive (or negative) associations with candidate names, etc. If an owner permits a forensics analysis that would develop only the collective, societal data about an OLSN, are such data subject to the same ethics considerations as individual data? Does an OLSN have privacy rights?

6 Privacy and Online Social Networks

6.1 Privacy

Prior research found that it is difficult to conceptualize privacy, perhaps because the concept is evolving over time. Privacy has not been well defined even though it is one of the fundamental values in society [35]. Thompson [36] notes that no one has a clear idea about the meaning of privacy rights. Prior literature does not even provide an accepted definition of privacy [37]. Researchers argue that setting up the parameters of privacy and the arguments for its protection is not easy even after a century in the development of privacy rights [38, 39].

The rights of data privacy are founded on the protection of individual rights against interference with personal decisions, against illegal searches, against intrusion on seclusion, unwanted spying, etc. Invasion of privacy is a tort arising from the exposure of damaging information regarding an individual [40]. Privacy rights protect against "the demands of a curious and intrusive society" [41, p. 958]. Fundamentally, an autonomous individual in the presence of a community is interdependent with that community. It is the norms of the community that enable the individual to exercise their autonomy. Privacy helps provide personal immunity and enables an individual to possess personality, autonomy, and thereby human dignity [41].

Privacy conventions and guidelines are implemented in myriad national, state, and provincial laws. Many countries have implemented omnibus national privacy laws that provide sweeping implementations of the conventions and guidelines, e.g. the European and Commonwealth nations. In others, notably the USA, there is more complex and diverse protection for privacy rights through a hodgepodge of special purpose laws. Such laws discretely protect the privacy of data related to health, education, finance, etc. [42].

Societies have tended to regard the growth in the power of information technologies for processing and communicating data as particularly threatening to the privacy of individuals' data. The owners of IT systems that store and process these data, and the owners of communications networks that transfer these data, are largely regarded as data stewards. Whether government- or commercially-owned, the duty to act as responsible stewards for personal data has become both a legal and social obligation [43]. Concerns over the forensics analysis of online social networks may further increase the tension between privacy and IT.

6.2   Online Social Networks

People in all cultures form complex social networks. Such networks can be as formal as a college fraternity or as informal as a neighborhood garden club. Such networks typically share values and trust. OLSNs are different from other communications media, and could be viewed as social milieux rather than media. OLSNs are information and communications technology (ICT)-based meeting places where individuals gather to socialize, that is, to exchange information, observe and emulate each other, and compare status [44].

OLSNs are not simply Web 2.0 implementations of off-line, real, or physical social networks using ICT. OLSNs are quite different. For example, these are functionally different. Unlike off-line, real, or physical social networks, OLSNs are particularly good for high speed, fast breaking communication and rapid socialization [45]. Further, OLSNs can actually present social structures that are opposite from or inverted from off-line networks. For example research has shown that individuals with lower self-esteem gain significantly more from participation in OLSNs than individuals with higher self-esteem [46].

Another difference that is particularly relevant to privacy is that an individual's online identity is not necessarily the same as the individual's off-line identity. Members of OLSNs may also operate on their own identities in rather instrumental ways. In other words some individuals create different identities for themselves in their OLSNs [47]. Members of OLSNs sometimes customize their identities to achieve their social goals [48].

Concerns for the individual privacy of members of OLSNs found voice almost from the beginning of OLSNs. The lack of privacy in these networks was obviously noticed as potentially harmful for those individuals who posted personal information in their profiles online [49]. It is important to recognize that one distinguishing characteristic of OLSNs is the vulnerability of data in electronic media to computer-based storage and analysis. A face-to-face interpersonal social exchange is characterized by its nature as ephemera; it has a brief and transitory existence. Because an online interpersonal social exchange is brokered by computers, the data can be subject to detective-style matching, analyzed on the fly, or trapped and stored for possible later analysis. Unlike a face-to-face exchange, an online interpersonal social exchange can be characterized by its nature as perpetual; it has a permanent and indefinite existence. As a result, OLSNs can be easily subjected to forensic study and investigation.

This point is not lost on the research community. Researchers are already demonstrating successful techniques for detecting and analyzing group characteristics and behavior. For example, research has helped us to recognize how OLSN members install cues in their public profiles to indicate their interests and motives [50]. Analysis of social network size has been shown to indicate a subject's gender and characteristics of introvert or extrovert behavior. In addition, large networks and extensive time spent online can distinguish opinion seekers from opinion leaders [51].

These research studies have the advantage of being able to easily and electronically capture and study the empirical evidence. However, these studies also illustrate how such methods also compromise sensitive data and expose the characteristics and behavior of an OLSN to open scrutiny.

For example, in one study seeking to develop better suicide prevention among youngsters, researchers have used automated data collection programs to distinguish young lesbian, gay, and bisexual individuals participating in an OLSN, and to study the structure of their networks [52]. While these purposes are beneficent, the same techniques could be applied with hostile intentions. Even the benign research is turning up surprising relationships from the analysis of OLSNs. For example, members who provide a religious affiliation in their public profile are also likely to install cues that indicate their interest in finding a romantic partner [50].

Here our study proposes the following research agenda and adds our discussion with regard to the privacy of OLSNs.

- In general, does a right to group privacy exist? If it does, what is the relationship between individual privacy and group privacy?
- With particular reference to online groups, do the privacy rights of individual members of an OLSN extend to the OLSN group itself? Does an OLSN as a group have the right to privacy?

6.3   Group Privacy

Much of the work on privacy laws has been focused on the individual right to privacy. This work refers to an individual's right to be let alone [53]. The notion of group privacy is more recent. Group privacy can be thought of as "the right to huddle" [54, p. 121]. While group privacy may lack a substantial theoretical foundation, it can be regarded as an extension of individual privacy to an association of individuals. Protection of group privacy is tightly linked to the duration, purpose, and context of the association.

The orientation of privacy concerns has always been individual. Expectations and security of group privacy is not pronounced. There appear to be three perspectives on the privacy rights of social groups. The first perspective holds that no group privacy exists. People with this perspective argue that there is no regulation to support group privacy. The second perspective holds that the privacy rights of the social group are nothing more than a collective of the privacy rights of the individual persons who belong to the social group. The third perspective would hold that the social group itself possesses an identity that is distinct from those of its individual members. Because the social group has a distinctive identity, it accrues a right to privacy that is distinctive from the simple collection of the privacy rights of its individual members.

The first perspective is reflected in current legal systems where many countries do not define or protect group privacy. In such legal paradigms, online social group behavior might be considered more or less public and can be investigated by anyone with access to both the media and the forensics technology, and to the legal degree that the group behavior cannot be traced to the individuals. In terms of whether group right-to-privacy itself exists as an ethical issue, some viewpoints may hold that a forensic analysis of online social groups does not violate anyone's privacy. Such a viewpoint is anchored to the lack of regulation as negating the existence of group privacy. An alternative view would acknowledge and protect of group privacy because it is an ethical issue and merely a legal oversight in current legal paradigms. Such an alternative view may be made timely by the increasing present of Internet-enabled OLSNs in the presence of network forensics capabilities.

The second perspective views that a social group has no distinctive identity apart from the collection of its members' identities. Therefore, individual privacy rights define any limits on the privacy of an online social group. Under this perspective, a compromise of group privacy means a compromise of individual privacy as the compromise can be traced to a single individual person, It is arguable that an online social group can be investigated without concern for privacy as long as the facts, behaviors, and characteristics making up evidence about the group cannot be traced to a single individual person.

The third perspective views that a social group has a distinctive identity apart from the collection of its members' identities, it may be possible to violate the privacy of the social group without necessarily penetrating the individual privacy of its members. In such a case it would be possible to collect evidence from within a social group that regards that group without compromising the privacy of any individual member of that group. Group privacy is a form of privacy that people seek in their association with others. As an extension of individual privacy, group privacy protects the desire and need of people to come together, exchange information, share feelings and make plans. Group privacy can be compromised without affecting individual privacy [55]. For example, group privacy is compromised when a newspaper reports a secret society's rituals that the society wishes to keep from the public eye. In this case, the group would experience loss independently of any loss of individual privacy.

If it is possible to collect evidence about a social group in such a way that it does not compromise the privacy of individuals, then the rights to investigate social groups may be poorly defined as well as protected in current law. There is little effective legislative or judicial restriction on forms of interference or invasion against the freedom of association [54]. This lack of legal protection further complicates the collection of evidence about online social groups because many online social groups are transnational in their membership. It appears to be the present case that anyone in the world has the right to collect evidence and investigate online social groups as long as data about the individual members is not compromised.

7   Ethics and Online Social Network Privacy

Whether the violation of OLSN's group privacy would lead to the violation of individual privacy is primarily an ethical issue because the regulation to protect group privacy does not exist. The complicating factor for the forensic examination of OLSN is the need to investigate the association without violating the privacy of members of the association who are not subjects of the investigation. It may be possible to forensically develop evidence about the association without collecting specific evidence about all members of the association. As a particular form of association of individuals, there may be precedence for considering OLSN privacy as distinct from the privacy of the individuals who make up the OLSN membership.

Two views (objective and sociological views) on the violation of an individual privacy exist when an individual's association with the group is revealed. Objectively, we know that OLSNs might have members with different opinions and views. When the group privacy of OLSNs is compromised in a way that the group's characteristics and the identities of its members are revealed, it should mean that members' individual privacy is compromised. Some members may not share values and opinions of groups. From this objective perspective of reality, still some harm might be done to those individuals' privacy through the disclosure of group characteristics and activities. Inaccurate information about individuals might be revealed in that case.

Sociologically, however, people assume the homogeneity of groups in a way that implies the group members share the same values and beliefs. They equate individual characteristics and behavior with the group characteristics and behavior (i.e., stereotyping). Even though this assumption may not hold in fact, we may regard it as sufficiently effective from a social constructionist viewpoint to be harmful to some individuals in the group [43]. Therefore in a very real way, when group privacy is compromised, individual privacy is also compromised. This argument means that group privacy should be protected as a collective sum of individual privacy. This view is based on the presumption of the common overestimation of the homogeneity of groups.

In addition to the issue of the relationship between group privacy and individual privacy, we need to check the stage of ethical reasoning with regard to the protection of OLSN's privacy. Ethical reasoning stages can be grouped into three levels: pre-conventional, conventional, and the post-conventional. The pre-conventional level is self-oriented and involves obedience and instrumental egoism (self-serving). The conventional level is oriented toward relationships and involves conforming to the expectations of others (Law and social order). The post-conventional level is been more oriented toward socially accepted and pronounced principles [56]. The current stage of OLSN's privacy protection is the stage of self-serving. The definition and protection of OLSN's privacy is left to forensic investigators' responsibility as no legal regulation has been set up.

A values perspective on ethical reasoning may involve four categories: egoism, utilitarianism, rule deontology, and social relativism [57]. This simple values framework helps to organize the conflicting ethical reasoning issues created by our growing ability to perform forensic analysis on OLSNs. Two perspectives depending on the membership of online social network can affect the interpretation of the four categories of ethical reasoning. A critical tension may develop between the online social group membership and the non-membership. This tension is typical whenever a social group exhibits the characteristic of exclusivity. The perspectives of ethical reasoning are largely relative to these two dimensions. Further research is necessary to investigate the relationship between the two frameworks: ethical reasoning values and perspectives of OLSN membership.

8 Discussion

The protection of group privacy in OLSN groups raises multiple issues. First, group privacy is integral for the freedom to organize and participate in social groups. This freedom is the foundation of democratic society. With the introduction of information technologies, an OLSN grows its own identity (e.g., group name, membership parameters, etc.). As a result, it can have its own privacy on a group level. Second, this OLSN group privacy can be compromised by an analysis using network forensics. The forensics techniques introduced above can also easily identify and reveal membership information about each individual. The knowledge of this can lead to members' apprehension over the revelation of their individual information.

The existing tension between network forensics and accessibility to OLSNs needs to be acknowledged and balanced in keeping with the benefits of the societies involved. Forensics techniques have a capability to identify and reveal information about individuals within OLSNs. As the knowledge about the capability of network forensics spreads via media to societies, this can make people reluctant to join and to be active in OLSNs. This reluctance might have negative impacts on societies' needs to express and share diverse opinions thereby integrating societies harmoniously. Further, network forensics is necessary to protect social justice. Network forensics can be both a threat and a benefit to society. The challenge rises in how to use network forensics for the benefits of society while minimizing its inherent risks.

Most of the current regulations and laws are aimed at protecting individual privacy. Group privacy also has its own cultural/sociological limits. Nations that span different cultures have their own distinctive privacy regulations. Group privacy would protect characteristics and activities that can be different from the characteristics and activities of the individual members. A group can have an average age, a dominant gender, typical behaviors, and other characteristics. In a secret religious gathering, for example, an overall, collective response as a group to a leader's message at the meeting would only be considered confidential if the concept of group privacy is acknowledged. Individual members of the group may or may not exhibit the same characteristics as the group. Indeed, with factors such as the group response above, we cannot assume that all anonymous individuals share the same attitude to the message. Individual privacy cannot necessarily be saved through anonymity in settings where group privacy is compromised.

Because group characteristics and behaviors can be different from individual characteristics and behaviors, group privacy is not the same as the sum of the individual privacy. It is possible to compromise group characteristics without compromising any one individual's privacy. Because few privacy rights are attached to a group, OLSN groups are open to a risk of compromise to network forensics. Because many OLSN are international in scope, there is a risk for abuse across national borders. Network forensics techniques can easily extend across borders. There are possibilities for the creation of a new form of digital divide, one in which participation in OLSN is risky for individuals whose association with an OLSN group might be regarded as dangerous behavior by their local authorities. Such a digital divide would separate "haves", those for whom free participation in OLSN engenders little social risk; and "have nots", those for whom such participation is dangerous and therefore avoided.

The new capacity to forensically examine online social networks opens new vistas for data forensics. However, it confronts the forensics examiners with new ethical and legal issues. Does group evidence from a forensics analysis of an online social network violate the privacy rights of the individual members of the group, or the privacy rights of the group as a distinct entity, or neither, or both? Does the potent application of network forensics to OLSNs create difficult social boundaries for participation in such groups? Future research needed in a number of areas. These include the ability to distinguish group and individual evidence, the need for regulations for group privacy, and ethical guidelines for forensic examinations of online social networks. For example, when group privacy must be compromised, an ethically fair purpose for intrusion should be pronounced. The ethical value of such intrusions should be established and weighed against the costs. The justifiable compromise of group privacy should be independent from the loss of individual privacy.

9   Conclusion

A group's purpose can be compromised when group members are reluctant to join and be active due to the concern over identification and revelation of their individual privacy. At risk are the social benefits of online groups. Establishing a social network online makes it more feasible to use computer analysis to develop knowledge about the network group itself, over and above any compromise of its individual members. As a result, an online social network opens new avenues for automated forensics scrutiny and new problems for the protection of privacy in a digital world.

The field of data forensics has developed sophisticated methods and tools for the purpose of collection and analysis of digital information for the purpose of detecting specific human behavior and producing evidence regarding this behavior. There is great social value available from these tools, but also great social risk of a different kind than modern societies have encountered before. Care must be taken to balance the benefits and costs of both online social networking and network digital forensics.

References

1. McAfee, A., Brynjolfsson, E.: Big data: the management revolution. Harv. Bus. Rev. **90**(10), 60–69 (2012)
2. Garfinkel, S.L.: Digital forensics research: the next 10 years. Digit. Invest. **7**, S64–S73 (2010)
3. Huber, M., Mulazzani, M. Leithner, M., Schrittwieser, S., Wondracek, G., Weippl, E.: Social snapshots: digital forensics for online social networks. In: Proceedings of the 27th Annual Computer Security Applications Conference, pp. 113–122 (2011)
4. Zainudin, N.M., Merabti, M., Llewellyn-Jones, D.: A digital forensic investigation model and tool for online social networks. In: 12th Annual Postgraduate Symposium on Convergence of Telecommunications, Networking and Broadcasting (PGNet 2011), Liverpool, UK, pp. 27–28 (2011)
5. Cocking, D., van den Hoven, J., Timmermans, J.: Introduction: one thousand friends. Ethics Inf. Technol. **14**(3), 179–184 (2012)
6. Vallor, S.: Flourishing on Facebook: virtue friendship & new social media. Ethics Inf. Technol. **14**(3), 185–199 (2012)
7. Hull, G., Lipford, H.R., Latulipe, C.: Contextual gaps: privacy issues on Facebook. Ethics Inf. Technol. **13**(4), 289–302 (2011)
8. Zimmer, M.: "But the Data Is Already Public": on the ethics of research in Facebook. Ethics Inf. Technol. **12**(4), 313–325 (2010)
9. Lusoli, W., Compañó, R.: From security versus privacy to identity: an emerging concept for policy design? Info J. Policy Regul. Strategy Telecommun. Inf. Media **12**(6), 80–94 (2010)
10. Posey, C., Lowry, P.B., Roberts, T.L., Ellis, T.S.: Proposing the online community self-disclosure model: the case of working professionals in France and the U.K. who use online communities. Eur. J. Inf. Syst. **19**(2), 181–195 (2010)
11. Tow, W.N.-F.H., Dell, P., Venable, J.: Understanding information disclosure behaviour in australian Facebook users. J. Inf. Technol. **25**(2), 126–136 (2010)
12. Mishra, A.N., Anderson, C., Angst, C.M., Agarwal, R.: Electronic health records assimilation and physician identity evolution: an identity theory perspective. Inf. Syst. Res. **23**(3), 738–760, 844, 846 (2012)
13. Kumar, V., Mirchandani, R.: Winning with data: social media-increasing the ROI of social media marketing. MIT Sloan Manag. Rev. **54**(1), 55 (2012)
14. Dunn, B.J.: Best buy's ceo on learning to love social media. Harv. Bus. Rev. **88**, 43–48 (2010)
15. Edelman, D.C.: Branding in the digital age. Harv. Bus. Rev. **88**(12), 14–18 (2010)
16. Court, D., Elzinga, D., Mulder, S., Vetvik, O.J.: The consumer decision journey. McKinsey Q. **2**, 66–77 (2009)
17. Sarathy, R., Robertson, C.J.: Strategic and ethical considerations in managing digital privacy. J. Bus. Ethics **46**(2), 111–126 (2003)
18. Ashworth, L., Free, C.: Marketing dataveillance and digital privacy: using theories of justice to understand consumers' online privacy concerns. J. Bus. Ethics **67**(2), 107–123 (2006)
19. Calluzzo, V.J., Cante, C.J.: Ethics in information technology and software use. J. Bus. Ethics **51**(3), 301–312 (2004)
20. Boyd, D.M., Ellison, N.B.: Social network sites: definition, history, and scholarship. IEEE Eng. Manag. Rev. **38**(3), 16–31 (2010)
21. Boyd, D., Ellison, N.: Social network sites: definition, history, and scholarship. J. Comput. Mediat. Commun. Electron. Ed. **13**(1), 210–230 (2007)
22. Kane, G.C., Fichman, R.G., Gallaugher, J., Glaser, J.: Community relations 2.0. Harv. Bus. Rev. **87**(11), 45–50 (2009)
23. Boyd, D.: Friendster and publicly articulated social networking (2004)

24. Granovetter, M.: The strength of weak ties: a network theory revisited. Sociol. Theor. **1**, 201–233 (1983)
25. Gross, R., Acquisti, A., Heinz III, H.: Information revelation and privacy in online social networks. In: ACM Workshop on Privacy in the Electronic Society, pp. 71–80. ACM, New York (2005)
26. Merriam-Webster: Forensics. In: Merriam-Webster Online Dictionary (2010)
27. Casey, E.: Digital Evidence and Computer Crime—Forensic Science, Computers and the Internet, 2nd edn. Academic Press, San Diego (2004)
28. Endicott-Popovsky, B., Frincke, D.A., Taylor, C.A.: A theoretical framework for organizational network forensic readiness. J. Comput. **2**(3), 1–11 (2007)
29. Anson, S., Bunting, S.: Mastering Windows Network Forensics and Investigation. Sybex, New York (2007)
30. Fuentes, F., Kar, D.C.: Ethercal vs. Tcpdump: a comparative study on packet sniffing tools for educational purpose. J. Comput. Sci. Coll. **20**(4), 169–176 (2005)
31. Conti, G., Dean, E.: Visual forensic analysis and reverse engineering of binary data. Black Hat USA (2008)
32. Kim, B.-K., Jang, J.-S., Chung, T.M.: Design of network security control system for cooperative intrusion detection. In: Chong, I. (ed.) ICOIN 2002, Part II. LNCS, vol. 2344, pp. 389–398. Springer, Heidelberg (2002)
33. Haggerty, J., Taylor, M., Gresty, D.: Determining culpability in investigations of malicious e Mail dissemination within the organisation. Presented at the Third International Annual Workshop on Digital Forensics and Incident Analysis, WDFIA 2008 (2008)
34. Stolfo, S., Creamer, G., Hershkop, S.: A temporal based forensic analysis of electronic communication. In: Proceedings of the 2006 International Conference on Digital Government Research, pp. 23–24. ACM (2006)
35. Westin, A.F., Blom-Cooper, L.: Privacy and Freedom. Atheneum, New York (1970)
36. Thomson, J.J.: The right to privacy. Philos. Public Aff. **4**(4), 295–314 (1975)
37. Inness, J.C.: Privacy, Intimacy, and Isolation. Oxford University Press, Oxford (1996)
38. Borna, S., Sharma, D.: Considering privacy as a public good and its policy ramifications for business organizations. Bus. Soc. Rev. **116**(3), 331–353 (2011)
39. Regan, P.M.: Legislating Privacy: Technology, Social Values, and Public Policy. University of North Carolina Press, London (1995)
40. Volokh, E.: Personalization and privacy. Commun. ACM **43**(8), 84–88 (2000). Association for Computing Machinery
41. Post, R.C.: The social foundations of privacy: community and self in the common law tort. Calif. Law Rev. **77**(5), 957–1010 (1989)
42. Baumer, D.L., Earp, J.B., Poindexter, J.C.: Internet privacy law: a comparison between the United States and the European Union. Comput. Secur. **23**(5), 400–412 (2004)
43. Shapiro, B., Baker, C.R.: Information technology and the social construction of information privacy. J. Account. Public Policy **20**(4), 295–322 (2001)
44. Clemons, E.: The complex problem of monetizing virtual electronic social networks. Decis. Support Syst. **48**(1), 46–56 (2009)
45. Farhi, P.: The Twitter explosion. Am. J. Rev. **31**(3), 26–31 (2009)
46. Steinfield, C., Ellison, N.B., Lampe, C.: Social capital, self-esteem, and use of online social network sites: a longitudinal analysis. J. Appl. Dev. Psychol. **29**(6), 434–445 (2008)
47. Howard, B.: Analyzing online social networks. Commun. ACM **51**(11), 14–16 (2008). Association for Computing Machinery
48. Young, K.: Online social networking: an Australian perspective. Int. J. Emerg. Technol. Soc. **7**(1), 39–57 (2009)

49. Rosenblum, D.: What anyone can know: the privacy risks of social networking sites. IEEE Secur. Priv. **5**(3), 40–49 (2007)
50. Young, S., Dutta, D., Dommety, G.: Extrapolating psychological insights from Facebook profiles: a study of religion and relationship status. CyberPsychol. Behav. **12**(3), 347–350 (2009)
51. Acar, A.S., Polonsky, M.: Online social networks and insights into marketing communications. J. Internet Commer. **6**(4), 55–72 (2007)
52. Silenzio, V.M.B., Duberstein, P.R., Tang, W., Lu, N., Tu, X., Homan, C.M.: Connecting the invisible dots: reaching lesbian, gay, and bisexual adolescents and young adults at risk for suicide through online social networks. Soc. Sci. Med. **69**(3), 469–474 (2009)
53. Warren, S.D., Brandeis, L.D.: The right to privacy. Harv. Law Rev. **4**(5), 193–220 (1890)
54. Bloustein, E.: Individual and Group Privacy. Transaction Publishers, Pallone (2002)
55. Dumsday, T.: Group privacy and government surveillance of religious services. Monist **91**(1), 170–186 (2008)
56. Myyry, L., Siponen, M., Pahnila, S., Vartiainen, T., Vance, A.: What levels of moral reasoning and values explain adherence to information security rules? An empirical study. Eur. J. Inf. Syst. **18**(2), 126–139 (2009)
57. Zwass, V.: Ethical issues in information systems. In: Drake, M.A. (ed.) Encyclopedia of Library and Information Science, pp. 1054–1062. Marcel Dekker, New York (2003)

Digital Leadership Through Service Computing: Agility Driven by Interconnected System and Business Architectures

Mohan Tanniru[1(✉)] and Jiban Khuntia[2]

[1] Oakland University, Rochester, MI 48309, USA
tanniru@oakland.edu
[2] University of Colorado Denver, Denver, CO 80202, USA
jiban.khuntia@ucdenver.edu

Abstract. Digitization to support services is an evolving phenomenon in a service-driven economy. While initiation and infusion of digitization in services may be relatively easy, the appropriation of value from such digitization is difficult. The digitization of services, represented as service objects, has to be modular and configurable to support business agility in order to assess viability and create value. While modularization of service objects is well established in existing literature, its ability to support business agility is not explored in the literature. In this study, we propose a concept of digital leadership that links service objects, represented in a service system architecture designed to support digital services, with various components of a business architecture for business agility in a changing business and technology landscape. We provide a definition of service objects to represent digital services, its associated characteristics, and how these service objects contribute to business agility. The concept is applied to a patient room digitization case operationalized under the theme of a service robot.

Keywords: Digital leadership · Service computing · Service robot · Agility · Robot

1 Introduction

With the growth of the service industry, service delivery is emerging as a new paradigm for value creation. Firms are leveraging information technology (IT) in the service rendering and delivery processes. The infusion of IT is leading to new business models, disrupting the existing brick-and-mortar paradigm and leading businesses to reach out to a global customer base [1, 2]. This emerging perspective and practice, which places services rather than products at the focus of business activity, is conceptualized as service dominant logic (SD logic). In this logic, firms need to coordinate many of their internal, customer and partner resources (referred to as operant resources in SD logic) to render customer-oriented services [24]. The firm's operating resources can be mostly the IT architecture that connects operant resources. For example, the world's largest car service company Uber does not own cars, the largest accommodation service company Airbnb has no hotel infrastructure, the largest movie house Netflix owns no theaters, and new

© Springer International Publishing Switzerland 2016
V. Sugumaran et al. (Eds.): WEB 2015, LNBIP 258, pp. 112–125, 2016.
DOI: 10.1007/978-3-319-45408-5_10

retail service houses have no inventories (e.g., Alibaba and Amazon). These firms are leveraging on operant resources in the SD logic, through a service-driven IT architecture, to create value.

The design, development and implementation of a solid service-driven IT architecture to create value is a complex process. However, this complexity may bring success or failure to the IT-enabled service business models. With increasing acceptance of the SD logic and value co-creation with customers, it is expected that value propositions will continue to change with customer experiences. A firm has to remain operationally flexible and leverage IT advances. It needs to develop basic, peripheral, and supporting services [4, 5] to meet changing customer value propositions and use IT advances (digitization) to improve the efficiency and effectiveness of processes within these services [1, 6]. In addition, it may also use external resources through service exchanges with other firms and customers [1]. While digitized services within a firm are aimed at achieving higher efficiency, service exchanges are used to help address evolving customer support needs. However, extant literature on market-focused IT-enabled service orientation or architecture is less mature [3].

Prior studies have argued that service firms need to embed IT in customer-oriented processes to achieve better outcomes [15, 16], and service delivery processes to address customer expectations [17] and influence firm performance [18, 19]. The changing technology landscape and the role of the customer in co-creating value is calling on businesses to align their business architecture (comprising various strategies used to address business objectives) with the service system architecture (comprising the way digitized services are creating customer value) – see Fig. 1. Three fundamental drivers are influencing the need for such an alignment: (1) the technology landscape, influenced by advanced digitization opportunities, is creating digitized services, (2) the customer landscape, influenced by digitized services, is seeking new value propositions, and (3) the competitive landscape, influenced by opportunities for differentiation, is looking for new business models.

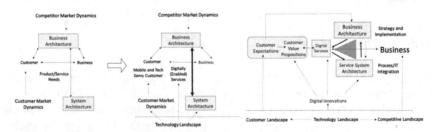

Fig. 1. Changing technology landscape and service enabling business architecture

IT-enabled service orientation is shown to support market differentiation by meeting changing customer expectations [6, 10, 20, 21] and enhancing business value [18, 19, 22, 23]. IT-enabled service strategies are shown to impact a customer's perceptions of a firm [9], effecting service quality and customer outcomes [10, 11], and help a firm exploit opportunities to compete successfully [12, 13] with augmented context based service-rendering activities [14]. However, there is a gap between the design, development and implementation of modular system architectures to support services and the business strategies that need to evolve to meet changing customer value propositions and sustain firm level performance [7, 8]. This gap is widely recognized both in practice and by academics alike.

This study addresses this gap by proactively linking the service system architecture with the business architecture, under the concept of digital leadership. The service objects of the service system architecture connect the digital services that create customer value with the strategies used in the business architecture to realize this value. The service objects thus help operationalize the service dominant (SD) logic. As value propositions change, service objects used to support the value propositions may be re-designed or configured, and associated operant resource adjusted and business strategies realigned, all to support business agility.

The next section provides the 10 steps associated with digital leadership. Section 3 uses the patient room robot as an example to discuss the first five steps of the digital leadership – innovation to system architecture. Section 4 shows how the modularity embedded in the service objects can help support business architecture (steps 6–10). Section 5 provides some concluding comments.

2   Digital Leadership

Digital Leadership is defined as the intersection of innovation, technology and strategy. It is not IT leadership, which aligns the delivery of digital products and services to meet business goals, or business leadership, which develops new products and services to meet changing customer needs, but an integration of business and IT leadership. It is service-driven, technology-enabled, and strategy-supported. More specifically, it aims to achieve three objectives. The first objective is to create and/or co-create customer value propositions that are supported through digital services by leveraging advanced information technology opportunities (digital innovations). This is illustrated as steps 1 and 2 in Fig. 2. The second objective is to develop a service system architecture, through service objects, that is highly modular and configurable. These service objects include modular digital products and data needed to deliver on performance metrics associated with digital services, and is illustrated by steps 3, 4, and 5 in Fig. 2. A third objective is to develop the business architecture, which includes strategies along multiple dimensions such as market positioning, competitive differentiation, partnership, risk

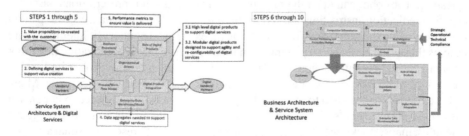

Fig. 2.   System and business architecture

mitigation, and implementation. These are intended to operationalize the service objects identified to implement digital services. This is shown in steps 6 through 10 of Fig. 2. The development of an agile service system architecture and associated business architecture can help a firm assess the technical and commercial viability of digital services respectively. This approach is illustrated using an innovation in a patient room: a patient room robot to support a patient's expectations regarding care delivery.

3 Patient Room Robot

To illustrate the digital leadership concept through the 10 steps discussed earlier, we conceptualize a patient room with a robot that acts as an integrator or coordinator of select services. The curtain raiser here is the imagination of a robot in a patient room. Let us imagine a patient at a hospital waiting for a procedure or recovering from an operation. A robot welcomes the patient to the room with a greeting, introduces itself and the staff that will be taking care of the patient, and turns on the patient's favorite TV station or begins playing music that he or she likes to watch. Interestingly, this may sound like a movie theater, but it is indeed a patient room in the not too distant future. The patient room has a bed and a chair, a toilet, two additional chairs connected with a flat wooden panel for family members to sit and relax when they visit the patient. A flat panel lifts up with a power outlet underneath if visitors need to charge their cellphone or computer. There is a pharmacy drawer in which to put medications, a security locker for valuables, and of course the personalized robot – that is the focus of this study. Three entities are related to the robot concept and are actors in this study: (1) the hospital employees, including doctors, managers, nurses, and support staff, (2) the patients, (3) a hypothetical firm P2R, which is developing the patient room robot. For the firm, the patients are the primary customers, the hospital is the service enabling and supporting platform, and the employees of the hospital are acting as agents in implementing the patient robot.

The robot knows why that the patient is there, and it can do a number of actions or service operations. First, the robot can turn the bed into a "smart bed" if the patient has a hip or leg procedure or is too weak to walk. The bed can then send signals to the nurse when the patient is trying to get up and possibly hurt himself or herself. The robot can call a nurse to give insulin if the patient is diabetic and orders food that requires an insulin tablet or injection. The robot can call staff to help collect all the necessary instrumentation when the patient has to be transferred to another unit for additional procedures or testing. The robot can open the pharmacy drawer and bring prescriptions based on specific medication instructions on the bottle and can call for assistance if the patient needs help using the toilet. Such robots do not yet exist, but a plethora of intelligent care systems that assist patients in a patient room today indicate that the realization of a robot is not far ahead. Laboratories and companies are developing interactive robot application and service platforms, such as VerBot, Xiao I robot, MSN robot, Chat robot etc. Many of these robots are designed to manage daily tasks and provide assistance to humans at a low cost. These robots can query and provide information on a number of items: schedules, maps, stocks, weather, television programs, etc. At present, personal assistant robots have emerged as a unifying theme for the integration of computation, communication, and interaction processes underlying the concept of assistance to the human.

The use of robots in health care is a very recent and useful concept [25–27]. In reality, practical implementations of personal assistance robots that can assist patients are sparse [28]. However, health care delivery within the walls of a hospital is highly person centric. As much as individuals need to adhere to some norms, face different alternatives, and subsequently have to choose from these alternatives, the burden of decision making and cognitive load is always high [29, 30]. These challenges aggravate when it has a bearing on a patient's life or death. For example, a nurse needs to attend to the sores of a patient stranded in a hospital bed, or needs to administer medications on time, and often may not be able to give personalized attention to another patient on a call. When a nurse can't address the needs quickly, the challenge remains on how to empower or enable a patient within his or her state to seek assistance from someone like a patient room robot.

What is missing today is a patient room is technology, such as a robot, to support communication and coordination. For example, the robot can tell staff as they come in or go out of the room to wash their hands by detecting, based on a badge code, if they have not done so already. It can also dispense bacteria-killing gel. The robot can introduce staff to the patient the first time they walk in and at other intervals so the patient can get to know them, turn on music or display on TV the artwork patients have seen in the hospital hallways, create the right atmosphere with music for informal interaction, change TV channels if needed, and ensure that educational material related to the patient's diagnosis is shown periodically (e.g., videos on hospital procedures, care details for when you leave the hospital, foods that can be consumed, prescriptions that should be taken, etc.). Perhaps most importantly, the robot can turn on the wrist-worn device that constantly monitors vital signals such as blood pressure, temperature and heart rate, and sends alerts to health care staff based on a wellness index uniquely calibrated to their condition.

To sum up, the robot concept in patient rooms is taking patient care to a new level, by integrating innovative patient room services with technology, and supporting the strategy of a hospital to address patient care quickly and effectively and improve their satisfaction. The purpose in this study is to make the robot configurable, so that the hospital, working with the patient and other care staff in a patient room, can co-create specific value propositions that are supported through a digitization of various patient room services. This patient robot design is then used to link the service and business architecture concepts discussed earlier.

3.1 Digital Service Innovations in the Patient Room

Identification of "digital service innovations", an intersection of technology and innovation, in a patient room is based on the work done by one of the hospitals (St. Joseph Mercy Health System in Oakland County, MI: SJMO). It is called an Intelligent Care System (ICS). The system was implemented in April 2014 in a 223-bed hospital and won the Wired Magazine's 2015 Innovator of the year award (See video at https://www.youtube.com/watch?v=dpnppt65oSs). Advanced digital technologies to support patient room care services are continually evolving. Recognizing opportunities and developing value propositions to support both changing patient expectations and address care-related needs (cost and quality) is a major challenge to health care organizations today. In this section, we will adapt some of the technologies embedded in patient rooms at SJMO and others to in the illustration of the patient room robot.

Many technologies–such as electronic badges for care delivery tracking, automatic distribution of education materials and drugs via electronic medical records, smart beds that track patient movement, patient call buttons to allow patients to request specific services (pain, bath room support, etc.)–have been widely discussed in the literature and are incorporated in the ICS [31–37]. Each of these technologies are embedded into the specific patient-care delivery staff interactions to provide value to the patient, as discussed below. The value propositions are then mapped to one or more digitized services (listed in Fig. 3).

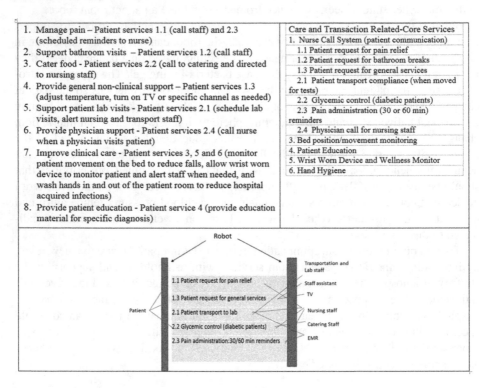

1. Manage pain – Patient services 1.1 (call staff) and 2.3 (scheduled reminders to nurse)	Care and Transaction Related-Core Services
	1. Nurse Call System (patient communication)
2. Support bathroom visits – Patient services 1.2 (call staff)	1.1 Patient request for pain relief
3. Cater food - Patient services 2.2 (call to catering and directed to nursing staff)	1.2 Patient request for bathroom breaks
	1.3 Patient request for general services
4. Provide general non-clinical support – Patient services 1.3 (adjust temperature, turn on TV or specific channel as needed)	2.1 Patient transport compliance (when moved for tests)
5. Support patient lab visits - Patient services 2.1 (schedule lab visits, alert nursing and transport staff)	2.2 Glycemic control (diabetic patients)
	2.3 Pain administration (30 or 60 min) reminders
6. Provide physician support - Patient services 2.4 (call nurse when a physician visits patient)	2.4 Physician call for nursing staff
7. Improve clinical care - Patient services 3, 5 and 6 (monitor patient movement on the bed to reduce falls, allow wrist worn device to monitor patient and alert staff when needed, and wash hands in and out of the patient room to reduce hospital acquired infections)	3. Bed position/movement monitoring
	4. Patient Education
	5. Wrist Worn Device and Wellness Monitor
	6. Hand Hygiene
8. Provide patient education - Patient service 4 (provide education material for specific diagnosis)	

Robot

1.1 Patient request for pain relief

1.3 Patient request for general services

Patient

2.1 Patient transport to lab

2.2 Glycemic control (diabetic patients)

2.3 Pain administration:30/60 min reminders

Transportation and Lab staff

Staff assistant

TV

Nursing staff

Catering Staff

EMR

Fig. 3. The patient services incorporated into the intelligent care system

3.2 Value Propositions

While many of the services listed below can be supported by a robot, for the purpose of this study we will focus on four value propositions, e.g., provide general non-clinical support, cater food, manage pain and support patient lab visits. The hypothetical firm P2R is engaged to develop the robot to digitize services associated with these four value propositions. The primary customers for all these services are patients in a hospital room, waiting for care or post-treatment during recovery and prior to discharge. The digitization of services to support each value proposition to the patient is apparent. However, an integration of these services, coordinated through a robot in the room, has an additional value to the patient: to give the patient control over communication and focused attention to how and when a request is being processed.

Besides the patient, there are others who the robot will support. Digitization of each patient service request, if handled by a nurse, can lead to too many alerts and contribute to stress (referred to as technostress [37]). The robot, as an integrator of all patient requests, can either perform some non-clinical needs (e.g., provide general information related to the TV, change the temperature of the room, etc.) or direct them to non-nursing staff (e.g., requests for bathroom support), thus reducing stress on the nursing staff. The patient room robot, by attending to a patient constantly, can also be a source of comfort to family members and possibly keep them informed of the patient's status. In other words, the innovation here can add value to both the nursing staff and the patient's family. For the discussion here, we will limit ourselves to value propositions that support patients.

Value propositions developed in support of patients will continue to change, and not all of them require the same type of services. For example, some patients need TV services and diabetic services, while others do not. Also, services needed for a cancer patient will be different from services needed by a patient who underwent a heart by-pass surgery. Moreover, no robot designed today can support the needs of patients in the future. For example, technologies to detect bacteria in shoes worn by family members or staff entering a room may not be available today, but may be developed in the future. This means that robotic services in the future (e.g., reminding family and staff to remove shoes before entering the room) may need to be added later. Also, some patients may need a version of a patient room robot taken with them when they leave the hospital for select services (e.g., prescription reminders, monitoring of vital signs, follow-up physician appointments, diets, etc.). In summary, the service system architecture used to design the patient room robot has to be agile to address changing value propositions. The next section will look at the components associated with the service system architecture.

3.3   Service System Architecture

The service system architecture includes the steps shown in Fig. 4. Each of these steps are discussed below for a single service. A service here is defined as the entire cycle: from a request to its fulfillment, and it includes all processes/work-flows used by both people and machines to fulfill this service request. Such a service-based visualization allows the firm to assess both the patient expectation of each service and measures used to assess the degree to which these services are fulfilling these expectations.

As noted earlier, a patient value proposition may be supported by a single or by multiple services. A single service may also support multiple value propositions, such as a "patient bed movement alert service" that addresses a patient's need to go to the bathroom while also reducing potential injury to a patient. Considering lab visitation service, the steps would involve the following:

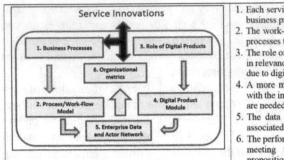

1. Each service request is translated into a set of required business processes
2. The work-flows currently used to execute the business processes to fulfill the service
3. The role of digitization in support of the service request, in relevance to the parts of the work flow that are altered due to digital innovation
4. A more modular approach to support the digitization, with the individual digital products (or applications) that are needed to support the service request
5. The data needed to support the digitization and the associated actor/data interface
6. The performance metrics used to evaluate the success of meeting the service needs and associated value propositions

Fig. 4. Steps in implementation of service system architecture

1. Support processes, such as recognizing a need to take a patient to a lab for tests, scheduling transportation, getting the patient ready to be picked up, and ensuring that the patient is returned to the room after lab tests are completed.
2. Set of work-flows used to execute each of these processes within the hospital. This may involve the sequencing of processes to fulfill the service request according to pre-established business rules or hospital protocols.
3. Advanced digitization opportunities that can be leveraged to improve the effectiveness or efficiency with which each service is fulfilled. In the example, a robot provides an integration of several of these work flows as shown in the Fig. 5.

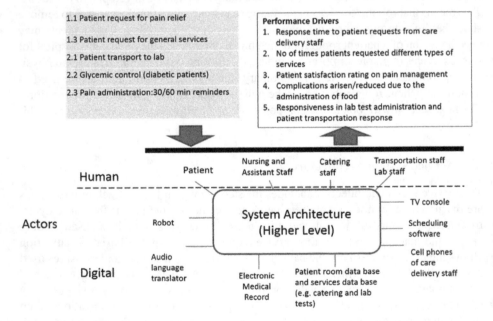

Fig. 5. High level the service system architecture for the robot

4. The role of the robot as an innovation is discussed in aggregate in step 3. For example, the monitoring of electronic medical records, confirming with the lab, and scheduling an appointment are all discussed in aggregate. However, in this step, this role is un-bundled or subdivided into high level digitization modules.
5. The data needed in the aggregate to support the digitization and the associated actor/ network relationships. The data here includes patient diagnosis and tests needs from electronic medical record, as well as specific patient room transaction data (such as patient call data related to tests, and lab tests data).
6. The performance metrics associated with the service: fulling the request related to the transportation of a patient to the lab. These metrics may include the number of times lab services were performed, patient satisfaction, and responsiveness.

In summary, the goal of the service system architecture is to operationalize the value proposition through various service objects. Each service object will identify the roles of multiple actors in fulfilling the service request, and each will have attributes such as: processes, work-flows, digital service modules that support these process/workflows, data and metrics. The Next section will discuss how this modularization will enable business agility.

4   Business Architecture

The P2R firm through its robot supports multiple patient room services using various service objects defined in the service system architecture (SSA). The ability of the firm to operationalize these digital services is evaluated along multiple dimensions of the business architecture: product positioning/marketing, partnership, risk mitigation, and implementation. While this a crucial step in assessing the commercial and operational viability of patient room robot, the goal in this study is to illustrate how a modular design of digital services via service objects can support the agility with which P2R can oper-ationalize its business.

Market Positioning: The envisioned robotic services may be marketed to hospitals for different reasons. A hospital may use these services to differentiate itself in a compet-itive health care market (current market, current product). Some hospitals may provide these services as an add-on to certain rooms with critically ill patients (e.g., patients post-surgery with high probability for complications) and reduce any associated costs (current market, new products). Others may offer these unique robotic services at a skilled nursing facility outside the hospital, in order to reduce patient readmission costs and improve patient satisfaction (new market, new service). The flexibility with which these robotics services can be offered will allow P2R to segment the hospital market and align its market positioning to address the strategic needs of different customer segments (hospitals).

Differentiation: The P2R firm can position its robotic services to support either cost or value based differentiation of a hospital. As discussed earlier, the metrics such as cost of care and patient satisfaction will help a hospital's need for focused cost leadership. One of the unique characteristics of a digital businesses is the intermediate value gener-ated by the data it creates. The data generated from the patient segments can be used by a hospital to tailor its services effectively and alter room assignments and staff allocation, and extend some of these services (diabetic support and lab support) post discharge. This can be valuable for hospitals to use robotic services for value based differentiation. A generalization of the robotic service data can also provide some unique differentiation and markets for P2R as well. For example, P2R generates data on patient services in a hospital room, and a synthesis of this data (in aggregate without violating patient rights) can lead to a product/service for an entirely new customer segment. Examples here include benchmark services that can reduce hospital costs, medical protocols for tests on patients with certain diagnoses, factors influencing patient satisfaction at a hospital, services that can be extended to patients at home or at a nursing home, or general services that can be provided to seniors or people with physical challenges. Some of these analyses may be offered as premium services for hospitals seeking differentiation strategy, or entirely to a new market (e.g. nursing homes). In other words, the modularity in the design of service objects not only helps P2R offer its robotic services to hospitals with cost or differentiation strategy, but also look for new market opportunities.

Partnerships: The patient room robot is an integrating device and needs access to communication protocols and data from a number of digital partners. For example, by reviewing the discussion in service system architecture, the robot has to be connected to a hospital's EMR and lab management, as well as staff scheduling systems (for scheduling transport staff). Also, the robot has to be linked to the nurse call systems that connect a patient to a nurse so the robot can act as an intermediary to reduce the number of alerts using additional intelligence. In other words, P2R needs transparent access to data and/or protocols for communication in order to integrate the robotic services with the hospital systems. At the same time, the robotics firm has to work with its own vendors such as language translation vendors (assumed as critical to support audio-based communication between patient and robot), and cloud-based vendors who will support web maintenance and some data management. Each vendor partnership has a differing impact on the success of P2R. The focus of the partnership can be cost-focused or value-driven (flexibility, configurability, reliability in decision making) as customer demands change.

Risk Mitigation: In a digital business, agility is key to changing value propositions. Agility to integrate digital services to meet changing customer expectations (e.g., language changes, changes in hospital operations), customize services for competitiveness (pricing, delivery, support, etc.), and work with partners (internal and at the customer site) to address technology and integration challenges all need an internal work-force that is adaptive and an organizational governance that is effective. There are several questions that should be asked when organizations undergo digital transformations. Is there the right talent within a firm and can the firm retain this talent? Do the roles and responsibilities of the people involved reflect the changing market? Is the change culture adding to security challenges and a firm's ability to protect its digital assets? Is the technology sufficiently mature to develop the product and integrate it at the customer site? Lack of readiness leads to risks, and digital businesses especially pose unique risks. These risks are classified by different authors as strategic, financial, operational, compliance, technical, reputational, etc. [38, 39]. Given the digitally intensive nature of the services, there is operational risk (includes talent related challenges), technical risk (includes both partner and cyber security challenges), compliance risk (includes governance and regulatory challenges), and strategic risk (includes financial, reputational and even partnership challenges). The ability of the P2R firm to form partnerships with other technology vendors can help mitigate technical risks, and having a right sourcing model (e.g., an outsourcing model for web platforms and strategic sourcing for language translation) can help mitigate operational risk.

Implementation: The P2R firm's implementation strategy is to be aligned with the speed with which a customer wants to realize the value from the new system. Is the hospital introducing the patient robot to every patient room, or only in certain rooms as a part of its initial diffusion strategy (i.e. radical vs. incremental change)? Is the health care delivery staff (nursing and health care personnel) sufficiently convinced and trained to use the technology, and are the hospital systems ready to integrate their systems with the robot? IT implementation of robotic services calls for data and process level integration with external stakeholders/systems along the boundary. For example, in patient room robot implementation, the boundary spanning systems (e.g., EMR, lab management, and catering) and stakeholders (physicians, lab technicians, and catering staff) have to adapt to changes dictated by the new patient room robot. Successful implementation strategies have to make these actors (people and systems) along the boundary see a value in such implementation. Can the EMR or lab management system gain valuable new information by tracking patient adherence to pain management and dietary intake? Can physicians and catering staff gain faster information on the clinical condition of a patient and/or food consumption patterns? Again, the degree of stakeholder commitment to adopt robotic services may dictate which robotic services will be implemented first and when. Again, the modularity in the service objects can help support varying implementation strategies.

In summary, the scope and scale of opportunities for service robots for patients is increasing and evolving. Patients may dictate the nature of digitized patient services that are demanded in the hospital and when they leave the hospitals. Patient rooms may exist not just in a hospital but in other places such as nursing homes, individual homes, etc. Such a wider lens may help the firm here to explore how it may scale its services to places other than a hospital. A patient may want certain services offered digitally no matter where the patient is, and some of these may not even be related to health care. For example, a patient sitting at home may want services such as guidance on pills for certain types of pain and types of food they can take when they are diabetic. Also, they may want to connect to a device virtually without knowing the details of the entertainment system or to speak to a device in a language that is not native to the system. Some of the services may be relevant even if a customer is not a patient by definition. Anyone "looking for information," and/or physically challenged to use traditional means to get that information, may need a service robot to provide such services. This may call for changing some of the robotic services designed, so they can be generalized for a broader market.

5 Conclusion

Digitization of hospital services in support of patient care is a competitive necessity for many hospitals seeking to reduce costs and improve patient satisfaction. This is both an opportunity and a challenge for firms providing digitized services. They need to understand both the changing hospital and health care market, regulations that dictate patient care, changing patient expectations, and volatility in various stakeholders with several hospital mergers and acquisitions. While patient room robots are used as an illustration to underscore the need for system and business agility, the abstraction of the domain (health care), the customer (patient or hospital), or the service (informational, transport, consumption, etc.) all call on digital service firms to explore opportunities to reach different market segments by making their services modular in nature. However, more importantly, each service object designed in the service system architecture (steps 3–5 with various attributes such as digital modules, data and service metrics) is mapped directly to the customer value proposition and digital services (steps 1 and 2) on the one hand, and business architecture (steps 6–10 with strategies on market positioning, value differentiation, partnerships, risk management and implementation) on the other. This makes the technology become tightly linked to the innovations to create customer value and the business strategy that can bring about this value. This is in fact what the digital leadership concept is: a way for both business and technology leaders to work in sync and, in iterative manner as appropriate, to ensure the propositions created and co-created are supported with agility in both service and business architectures.

References

1. Lusch, R., Vargo, S., Tanniru, M.: Service, value networks and learning. J. Acad. Mark. Sci. **38**, 19–31 (2010). doi:10.1007/s11747-008-0131-z
2. Vargo, S.L., Lusch, R.F.: Service-dominant logic: continuing the evolution. J. Acad. Mark. Sci. **36**, 1–10 (2008). Published online Epub Spring 2008
3. Aranda, D.A.: Service operations strategy, flexibility and performance in engineering consulting firms. Int. J. Oper. Prod. Manage. **23**, 1401–1421 (2003)
4. Butcher, K., Sparks, B., O'Callaghan, F.: Beyond core service. Psychol. Mark. **20**, 187–208 (2003). doi:10.1002/mar.10067
5. Anderson, S., Pearo, L.K., Widener, S.K.: Drivers of service satisfaction. J. Serv. Res. **10**, 365–381 (2008). doi:10.1177/1094670508314575
6. Menor, L.J., Roth, A.V.: New service development competence and performance: an empirical investigation in retail banking. Prod. Oper. Manage. **17**, 267–284 (2008). doi:10.3401/poms.1080.0034
7. Sambamurthy, V., Zmud, R.W.: Research commentary: the organizing logic for an enterprise's IT activities in the digital era—a prognosis of practice and a call for research. Inf. Syst. Res. **11**, 105–114 (2000)
8. Allen, B.R., Boynton, A.C.: Information architecture: in search of efficient flexibility. MIS Q. **15**, 435–445 (1991)
9. Barrutia, J.M., Gilsanz, A.: Electronic service quality and value: do consumer knowledge-related resources matter? J. Serv. Res. **16**, 231–246 (2012)

10. Ray, G., Muhanna, W.A., Barney, J.B.: Information technology and the performance of the customer service process: a resource-based analysis. MIS Q. **29**, 625–652 (2005)
11. Mithas, S., Krishnan, M.S., Fornell, C.: Why do customer relationship management applications affect customer satisfaction? J. Mark. **69**, 201–209 (2005)
12. Kathuria, A., Konsynski, B.R.: Juggling paradoxical strategies: the emergent role of IT capabilities. In: Thirty Third International Conference in Information Systems, Orlando, FL (2012)
13. Saldanha, T., Krishnan, M.: Leveraging IT for business innovation: does the role of the CIO matter? In: Thirty Second International Conference in Information Systems, Sanghai, China (2011)
14. Khuntia, J., Mithas, S., Agarwal, R., Roy, P.K.: Service augmentation and customer satisfaction: an analysis of cell phone services in base-of-the pyramid markets. In: Thirty Third International Conference in Information Systems, Orlando, FL (2012)
15. Damar, E., Hunnicutt, L.: Credit union membership and use of internet banking technology. BE J. Econ. Anal. Policy **10**, 1–30 (2010)
16. Allred, A.T., Addams, H.L.: Service quality at banks and credit unions: what do their customers say? Manag. Serv. Qual. **10**, 52–60 (2000)
17. Harvey, J., Lefebvre, L.A., Lefebvre, E.: Flexibility and technology in services: a conceptual model. Int. J. Oper. Prod. Manage. **17**, 29–45 (1997)
18. Byrd, T.A., Turner, D.E.: Measuring the flexibility of information technology infrastructure: exploratory analysis of a construct. J. Manage. Inf. Syst. **17**, 167–208 (2000). doi: 10.2307/40398473
19. Tafti, A., Mithas, S., Krishnan, M.: The effect of information technology-enabled flexibility on formation and market value of alliances. Manage. Sci. **59**, 207–225 (2013)
20. Grönroos, C.: Service Management and Marketing: A Customer Relationship Management Approach. Wiley, Chichester (2000)
21. Ives, B., Vitale, M.R.: After the sale: leveraging maintenance with information technology. MIS Q. **12**, 7–21 (1988)
22. Sambamurthy, V., Bharadwaj, A., Grover, V.: Shaping agility through digital options: reconceptualizing the role of information technology in contemporary firms. MIS Q. **27**, 237–263 (2003)
23. Saraf, N., Langdon, C.S., Gosain, S.: IS application capabilities and relational value in interfirm partnerships. Inf. Syst. Res. **18**, 320–339 (2007)
24. Rai, A., Sambamurthy, V.: Editorial notes-the growth of interest in services management: opportunities for information systems scholars. Inf. Syst. Res. **17**, 327–331 (2006)
25. Zelinsky, A.: Field and Service Robotics. Springer Science & Business Media, Heidelberg (2012)
26. Kang, J.W., Hong, H.S., Kim, B.S., Chung, M.J.: Work assistive mobile robots assisting the disabled in a real manufacturing environment. Int. J. Assist. Rob. Mechatron. **8**, 11–18 (2007)
27. Bemelmans, R., Gelderblom, G.J., Jonker, P., De Witte, L.: Socially assistive robots in elderly care: a systematic review into effects and effectiveness. J. Am. Med. Dir. Assoc. **13**, 114–120 (2012). Article ID e111
28. Moustris, G., Hiridis, S., Deliparaschos, K., Konstantinidis, K.: Evolution of autonomous and semi-autonomous robotic surgical systems: a review of the literature. Int. J. Med. Rob. Comput. Assist. Surg. **7**, 375–392 (2011)
29. Spath, P.L.: Error Reduction in Health Care: A Systems Approach to Improving Patient Safety. Wiley, San Francisco (2011)
30. McCarthy, M.: US hospital errors have fallen, says government report. BMJ **349**, g7461 (2014)

31. Galinato, J., Montie, M., Patak, L., Titler, M.: Perspectives of nurses and patients on call light technology. Comput. Inf. Nurs. **33**, 359–367 (2015)
32. Roszell, S., Jones, C.B., Lynn, M.R.: Call bell requests, call bell response time, and patient satisfaction. J. Nurs. Care Qual. **24**, 69–75 (2009)
33. Lasiter, S.: "The Button" initiating the patient–nurse interaction. Clin. Nurs. Res. **23**, 188–200 (2014)
34. Deitrick, L., Bokovoy, J., Stern, G., Panik, A.: Dance of the call bells: using ethnography to evaluate patient satisfaction with quality of care. J. Nurs. Care Qual. **21**, 316–324 (2006)
35. Klemets, J., Evjemo, T.E., Kristiansen, L.: Designing for redundancy: nurses experiences with the wireless nurse call system. Stud. Health Technol. Inform. **192**, 328–332 (2012)
36. Klemets, J., Evjemo, T.E.: Technology-mediated awareness: facilitating the handling of (un)wanted interruptions in a hospital setting. Int. J. Med. Inform. **83**, 670–682 (2014)
37. Khuntia, J., Tanniru, M., Weiner, J.: Juggling digitization and technostress: the case of alert Fatigues in the patient care system implementation. Health Policy Technol. **4**, 364–377 (2015)
38. Nissen, V., Marekfia, W.: The development of a data-centred conceptual reference model for strategic GRC-management. J. Serv. Sci. Manage. **7**, 63 (2014)
39. Sambamurthy, V., Zmud, R.W.: Arrangements for information technology governance: a theory of multiple contingencies. MIS Q. **23**, 261–290 (1999)

Are Online Reviewers Leaving?
Heterogeneity in Reviewing Behavior

Parastoo Samiei[(⊠)] and Arvind Tripathi

Business School, University of Auckland,
12 Grafton Road, Auckland 1010, New Zealand
{P.samiei, A.tripathi}@auckland.ac.nz

Abstract. Online consumption communities evolve over time and go through different stages in their life cycle [1]. The key factor of the sustainability of the community is members' ongoing contribution. This study examines the factors affecting the ongoing contribution of online reviewers for different types of users. Drawing from theory on communities of consumption [2] and popularity effect [3]; we propose a conceptual model of drivers of ongoing contribution. We observed that Social ties, sidedness, and consumption activity could explain the heterogeneity of ongoing contribution level for different users. We studied a community of book reviews. We showed that the effect of sidedness on contribution prediction is stronger for reviewers with extreme behavior. We also concluded that consumption activity has more predictive information about the contribution compare to the social tie and sidedness.

Keywords: eWOM · Social ties · Consumption activity · User type · Community of consumption · Sidedness

1 Introduction

Online consumption communities, like traditional communities evolve over time and go through different stages in their life cycle [1]. Although attracting new members in early stages such as creation and growth is important, retaining members during maturity stage is no less critical for their long-term success and sustainability. Communities are sustainable if they can maintain members' commitments and contributions over time. To achieve this goal, they develop mechanisms to encourage participants to maintain their contribution levels. These mechanisms will be more productive if online community detect and understand the drivers of members' ongoing contribution.

A community providing electronic Word-of-Mouth (eWOM) is a form on online community of consumption, which reduces the information search cost for potential customers. *Source credibility* is an important factor affecting the decision of the receiver/reader of a review to trust and adopt a review. On retailer sponsored, customer-to-customer review platforms, such as Amazon.com, most of the users do not know each other. The only information they have about the source credibility [4] is what is shared on the platform about the reviewer. In contrast, online communities of consumption shape around consumption of a product type (e.g., Goodreads.com for books), attract hobbyists and have stronger connections among members. In these

V. Sugumaran et al. (Eds.): WEB 2015, LNBIP 258, pp. 126–142, 2016.
DOI: 10.1007/978-3-319-45408-5_11

communities, with so many product experts and ongoing discussions on reviews and comments, it is not easy for members to write a review before consuming a product (e.g., read a book).

In an online community of consumption [2], members have some expertise or experience with the same product type. Even when the expertise is self- reported, it has an effect on the helpfulness and adoption of the review [5]. The experience and reviewer's history is usually available and eWOM receiver has the opportunity to explore them to decide if the reviewer shares a point of view or taste with her/him. Therefore, we argue that these communities offer more value to prospective customers compared to the reviews posted on retailer websites (e.g. Amazon), where a reviewer may have reviewed different kind of products, challenging their expertise in one particular area.

Online communities of consumption thrive on voluntary contribution of users and reviewers. Therefore, the continuous contribution of reviewers is vital for their long-term success [6]. However, many communities fail to attract or retain enough experts, experienced reviewers. Therefore, are unable to sustain in long run.

Reviewers have different characteristics and motivations, which can explain different behavioral patterns to some extent. Reviewing experience [7], user characteristics such as specialized skills [8], expertise [5], and motivation [9, 10] toward the social networks are some of those characteristics.

In this research, we intend to understand these behavioral patterns with the lens of reviewers' *ongoing contribution*. We aim to explore the factors that drive long-term ongoing contribution from reviewers on these communities of consumption [2]. Since reviewers are heterogeneous in their reviewing behavior, we propose a model drawing from the literature on e-tribalized marketing [2], to uncover the drivers of this heterogeneity. We concluded that *Social ties* and reported *consumption activity* are the main drivers of different behavioral patterns. Reviewers with stronger social ties may get popular on their community and a recent research showed that the popularity affects reviewing behavior including the sidedness of the reviews [3]. Following that, we argue that the valance of the reviews is likely to influence the reviewing behavior.

The paper is structured as follows: we begin with a brief overview of the relevant research develop our theory and hypotheses. We then describe our proposed model, followed by *analysis and results*. We conclude with discussions of our results and implications of our work.

2 Previous Work and Literature Review

Electronic Word-of-Mouth or eWOM is an informal information exchange in which former customers share their experience with prospective customers [11]. Current literature can be categorized in two main streams. **First stream** focuses on the generation of eWOM and concentrates on why and how customer share their reviews on online platforms [12]. Studies on the incentives and motivation of reviewers [13–15] and the content of reviews [16, 17] are some examples. The **second stream** focuses on the consequence of eWOM. Some examples are the effect of eWOM on sale [18–20]. In this area some researchers focused on the effect of reviews on the prospective.

Customers and the social interactions between reviewers and customers [3, 21]. eWOM literature has also been summarized from different perspectives. From the Products/Services Adoption Process perspective, Montazemi and Saremi [22] proposed a conceptual model in which they have categorized eWOM into five dimensions of *receiver, source, eWOM content, response, and focal product/service*. They have categorized the effect of these five dimensions on the three stages of product/service adoption. These five dimensions reflect the idea of Cheung and Thadani [4] in the literature review summary of eWOM. They have considered eWOM as a communication process, which includes a *sender, receiver, message,* and *response*. Later King et al. [23] consolidated the eWOM literature using a framework of interaction of two factors: *unit of analysis* and *cause-effect*. As the unit of analysis, they suggest sender and receiver. For addressing the cause and effect, they focused on the antecedence (as the cause) and consequence (as the effect) of eWOM.

From the current eWOM literature, we know that reviewers and participants in online communities of consumption are heterogeneous in their behavior [13, 24, 25]. Different factors shape this heterogeneity. Demographic attributes, personal characteristics, and motivations are some of these factors. Munzel et al. [13] summarized the motives of online reviewers for their contributions and studied different reviewing behaviors. They observed three distinct patterns of behavior, which resulted in three reviewer types: *creators, multipliers,* and *lurkers*. *Creators* are the users interested in first order activities such as reading and writing reviews. Whereas, *multipliers* are users with high passive activities. They are more interested in second order activities such as writing comments and re-sharing the content on the website. The last reviewer type is *lurkers* who are not interested in creating and sharing content. They are just interested in passive activities such as reading reviews [13]. Other researchers also found these different behavior patterns or user types for online reviewers. Lurking versus general participation [26, 27], lurkers versus posters [24, 25] are some examples of these studies. Some researchers also investigated the dynamics of different user types over time and examined their motivations. As an example, Preece and Shneiderman [28] suggested that reviewers and participants in eWOM communities change from a *reader* to *contributor*. If they engage in interaction with other reviewers, they become *collaborators* and if they continue to provide content (reviews) and share it with others; they are likely to become opinion *leaders* in the community.

In the current literature, we know about this behavior heterogeneity. However, we still cannot explain it. We try to focus on this gap and answer the question of what drives the heterogeneity in ongoing contribution of online reviewers.

3 Theory and Hypotheses Development

3.1 Online Community of Consumption

Internet facilitates the emergence of online communities. In online environment, millions of people form groups around different subjects. They share information, develop group identities and engage in social interaction with others. With the wide spread of creating and using eWOM, many of these communities are structured around consumption experiences [2]. Sometimes, a brand or a theme is the center of the community. (i.e. Star Treck [29]). Some other communities form around a specific product type (i.e. Goodreads for books, IMDB for movies, CNET for tech products, tripadvisor for travel, and BoardGameGeek for board games).

There are countless number of online communities out there. However, most of them are not sustainable. They may fail to engage enough expert members or experienced reviewers in the first place. They also may fail to keep their members motivated to continue their contribution and loose them over time. Therefore, sustainability of an online community is an indicator of the success of a community [30]. Concluded from previous works, Butler believed that sustainable community should have access to the pool of resources to support the social processes. He argued that a community with access to enough resource can provide the promised benefits to its member over time in exchange for the membership cost, and whoever stays longer will get these promised benefits [31]. Maintaining membership can lead to the commitment to the community. Members are constantly evaluating the community to check if it is rewarding enough to continue. This evaluation is affected by their commitment and it determines the likelihood of remaining in the community with the same level of contribution [31].

In the maturity stage of a community, poor participation, especially from members with weak ties can be the initiator of the dead stage of the community [1]. Therefore, it is vital for online communities to know their members and their behavior patterns, and understand the triggers of these patterns. They can use this understanding in designing mechanisms to maintain members and increase their commitment and contribution to avoid the dead stage.

3.2 Contribution and Continuity

In an online community of consumption, a reviewer can stop posting content without even informing the platform. Many reviewers gradually lower their level of contribution and become inactive. That shows the importance of understanding reviewers' behavior and their contribution for review hosting websites. As mentioned before, this research aims to study factors that drive long-term and ongoing contribution on an online consumption community.

Over time, reviewers gain experience, learn, and change their reviewing behavior and level of their contribution [7]. However, existing research is not conclusive about the direction of this change. A few studies have found that time, to be exact, the length of the membership; have an effect on reviewers' contributions through two mechanisms [32]. First, we expect the contribution to drop over time as the novelty and affection of the community decreases. Reviewers get bored or disappointed with the community and lower their participation. However, another mechanism works in very different way. Reviewers share their opinions with other people in the community and get feedback about the quality of their opinions. As rational agents, we expect them to get these feedbacks and modify their behavior hoping to get more and better feedback. So time, has an effect on reviewers' contribution based on their motivation and type of their activities [32].

3.3 Types of Consumption Community Members

To study the reviewer heterogeneity, we draw on the types of online community members in e-tribalized marketing. Kozinets [2] explained the identity formation of a member in an online community of consumption. We selected this model, because we believe that it is comprehensive enough to cover and explain different aspects of known reviewing behavior patterns we drew on from the literature as lurkers, posters, and multipliers, opinion leaders and followers and so on [24, 25, 28, 33].

Kozinets [2] categorized the contribution behavior of community members using two factors: *Consumption Activity* and *intensity of social relationships (social tie strength)*. Although these two factors affect one another, but they are separate enough that Kozinets [2] used them to categorize members based on their observed performance.

Consumption activity is the relationship of the user to the product and it shows how using that specific product is central and important in one's psychological self-image. This importance may increase the frequency of using the product. Assuming that writing reviews are consequence of use, the number of reviews written by an individual might reflect the importance of the product for her/him. On the other hand, *Social ties* represent the intensity and number of mutual relationships each user form with other members of the community. Kozinets [2] suggested that members belong to one of *tourist, Mingler, Devotee,* and *Insider* groups. *Tourists* are users with low consumption activity and lack strong social ties. They show a meager interest in the consumption (product) and their relationship to the community is minimum and casual. Many users act as Tourists when they join an online community and only engage in information browsing and lurking activities. If an online community grabs their attention, they are likely to transform to Minglers or Devotee. It depends whether they are more interested in the product or the social activities they engage in. If they like the product, better than their audience they will turn into a Devotee. Otherwise, they will become a Mingler who share fewer opinions, but develop a bigger social network.

On the other hand, if the social relation and information/knowledge sharing becomes interesting to Tourist, they are likely to engage in more relationships, develop strong social ties with others, and join Minglers who do not engage in product related. In general, it is possible for any of Minglers or Devotees to upgrade to Insiders if they thrive to maintain strong social ties and engage in high consumption activity (reviewing products in case of eWOM).

As aforementioned, Nov. et al. [32] concluded that the contribution of reviewers in online communities will decrease as the affection and excitement weaken over time unless users receive social feedback which keep them interested to the community. Therefore, we believe that if reviewers in an online community of consumption (eWOM platform) do not show interest in either social engagements or consumption activity (product) they are likely to lose interest and consequently their contribution level will drop even more. Hence, we hypothesize:

- **Hypothesis 1:** In an online community of consumption, members with weaker social ties and weaker consumption activity (Tourists) are more likely to have a *lower ongoing contribution* comparing to other members.

On the other hand, there is some research showing that the contribution level in the sense of number or reviews posted by one member is expected to decrease over time as they gain experience [34]. Reviewers may choose to spend more time to write a less number of reviews with higher quality. Despite this argument, we believe that the effect of the centrality of the consumption activity in one's life and reviewer interest in that specific product will lead to a different behavior.

Based on Kozinets [2], the consumption activity reflects the importance of the product for the reviewer. For example, an interest in technological Gadgets and electronic devices can be very close to heart for some reviewers on technical review hosting websites (i.e. cnet.com). Similarly, many members at Goodreads.com could be passionate readers because hobbyists created review community at Goodreads.com. We argue that reading books can be an important activity for many of these reviewers' on Goodreads.com. In that situation the importance of the consumption, activity trumps the effect of social network. These users stick to their ongoing contribution no matter what the social feedback they get is. To explore this effect, we hypothesize that:

- **Hypothesis 2:** In an online community of consumption, for members with the similar level of social ties, *level of ongoing contribution* will be driven by the levels of their consumption activity.

3.4 Social Network and Popularity Effect

The current literature on social network and its effect on eWOM can be categorized in two main streams. The first stream concentrates on the effect of the eWOM on the social network of the reviewers through the *Social Influence* concept [21, 35]. In this stream researcher focused on the fact that adopting reviews written by someone in one's social network during the pre-purchase or consuming phases significantly affects her/his product evaluation [36]. This effect can be bigger if the review writer is a friend [21]. The second stream is the effect that being in a social network has on the reviewer, and consequently, his/her reviewing behavior [3, 6, 7, 34].

In this research, we draw on Goes et al. [3] who also studied the popularity effect of the reviewer on their reviewing behavior. They showed that being popular in the online community of consumption (eWOM social network) increases the chance of getting feedback from other members and can drive an increase in both number of reviews and the number of negative reviews one member posts. It also affects the objectivity and readability of reviews. We draw on this effect of popularity on volume and valance (sidedness) of the reviews capture and explain the dynamics by which members in an online eWOM network change their behavior over time. (i.e. from being a Mingler or Devotee to an Insider and so on).

This result is in line with the social feedback mechanism of Nov et al. [32]. The size of the social network a reviewer is connected to indicate the size of her/his potential audience. Having a bigger audience can justify the cost and time needed to write a review [3] (assuming that there is some social motivation is in place). With this mechanism in place, we reckon that Insider reviewers, who have strong connections with other members, have a bigger audience and have more chance to get social feedback on their contributions. They also share more about their consumption experience and with more opinions out there, they may might more feedback. These feedbacks may result in increasing their contribution. Therefore, our hypothesis is:

- **Hypothesis 3:** In an online community of consumption, members with stronger social ties and higher consumption activity (Insiders) are more likely to have a *higher ongoing contribution* comparing to other reviewers.

As Goes et al. [3] also used a negative bias in social network to explain how popularity can systematically affect the product evaluation result and result in negative valance of reviews. This negative bias exists in social feedback, as people perceive that the negatively-valence review shows the unfavorable feelings and about the failure or the product in meeting reviewer's expectation. They usually perceive that reviewer is more knowledgeable or has higher standards [33]. This leads to our conceptual model of online customer reviewer behavior patterns in which we suggest that we need three axis to explain the different behavioral pattern of online reviewers (Fig. 1).

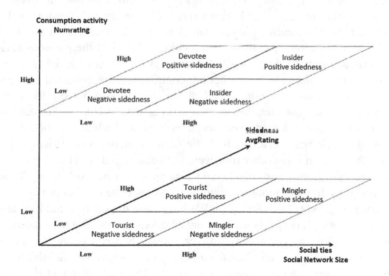

Fig. 1. The conceptual model of reviewer behavior patterns

- **Hypothesis 4:** In an online community of consumption, for members with similar consumption activity levels, *level of ongoing contribution* will be driven by the strength of their social ties.

Based on the proposed conceptual model (Fig. 1), we believe that sidedness can explain some of the heterogeneity of reviewers in their behavior patterns.

- **Hypothesis 5:** In an online community of consumption, members with negative overall sidedness are more likely to have a *higher ongoing contribution* comparing to other reviewers.

4   Analysis and Result

In order to answer our research questions, we have proposed a conceptual model (Fig. 1). In this model, we intend to study the interaction of *Sidedness, social tie strength,* and the *consumption activity.* We also investigated how these interactions can explain different behavior patterns of online reviewers.

We have selected books to control the effect of product type in our research. On the other hand, as books are considered experience goods. Experience goods versus search goods are products that a customer only can evaluate them after purchasing and using those [18] or by gathering information in other costly way. Online reviews can immensely decrease the information seeking cost for prospective readers (customers) and could have a huge effect on their purchase decision. Therefore, the selected online community has an important role in the market place for this product.

For data collection, we used an online crawlers and collected data from a eWOM social platform, which host online reviews on books. Our data set includes randomly selected 620 reviewers and their reviewing history. We collected data in several waves between June 2012 and July 2014. However, we have treated *Time* as a dynamic phenomenon. t_0 is when each reviewer started reviewing products on the website is different for each reviewer. t_1 is the time we collected data of the previous reviewing history, which is a specific date for all reviewers. As we wanted to investigate if the previous reviewing pattern affect the ongoing contribution, we collected data again on t_2 to calculate the measures of our dependent variable. t_0 is also a specific date.

We have calculated our measures of interest. For categorizing reviewers based on their historical behavior, we used three measures; first, as the indicator of the social tie strength, we have $SNSize_{ij}$ measure. It is the number of two-sided friendship bound formed between person i and other reviewers between t_0 and t_1. Then we categorized reviewers into two groups with bigger and smaller social network using its Median ($SNSizeLevel_{ij}$). As the consumption activity we have used the number of rated books by person i between t_0 and t_1 ($NumRating_{i1}$). We have used the same method to categorize reviewers ($NumRatingLevel_{i1}$). As the measure of sidedness, we also used the average of all rating scores given by person i to different books between t_0 and t_1 and used its median to categorized reviewers ($NumRatingLevel_{i1}$ and $AvgRatingSidedness_{i1}$).

Contribution level is our *dependent variable.* We Draw on Chen and Huang's work [37] in which they quantified contribution level using *review frequency* and *reviewing continuity.* We have defined $Continouity_{i2}$ as a binary variable with 1 valued if the reviewer rated any book between t_1 and t_2. In contrast, for *review frequency* we have defined a continuous variable showing low activity of the reviewers. To calculate that we have calculated the average days between each two reviews (which is a unique value for each reviewer based on their history. Then we calculated the expected activity of that reviewer assuming that they will continue to contribute with the same pace. Tables 1 and 2 include definition, calculation method, and descriptive statistics of all measures.

Table 1. Measures of interest

Variable	Definition	Calculation
$Continouity_{i2}$ (As a proxy for reviewing continuity)	It shows if person i rated any book between t_1 and t_2	1—if $(numrating_{i2} - numrating_{i1}) > 0$ 0—if $(numrating_{i2} - numrating_{i1}) = 0$
$LowActivityRatio_{i2}$ (As a proxy for review frequency)	The level of person i low activity book between t_1 and t_2, compared to his/her rating history	$= 1 - \frac{numratinig_{i2} - numratinig_{i1}}{expectedRatings_{i2}}$
$AverageDaysBetweenRatings_{i1}$	It is number of days on average person i post a new review, based on his/her rating history	$= \frac{numrating_{i1}}{(t_1 - t_0)}$
$ExpectedRatings_{i2}$	Number of ratings we expect done by person i between t_{i1} and t_{i2}, with his/her average behaviour at t_{i1}	$= \frac{(t_2 - t_1)}{AverageDaysBetweenRatings_{i1}}$
$AvgRating_{i1}$	The average of rating scores given by person i to different books between t_0 and t_2 (between 1 and 5)	collected from reviewers' profile
$AvgRatingSidedness_{i1}$	Is a binary value represents the average sidedness of evaluation for person i at t_1 compared to the median of sidedness of our dataset	For reviewer k 1—if $AvgRating_{k1} >$ Median $(AvgRating_{i1})$ 0— if $AvgRating_{k1} <$ Median $(AvgRating_{i1})$
$NumRating_{ij}$	The number of rated books by person i between t_0 and t_j	collected from reviewers' profile
$NumRatingLevel_{i1}$	Is a binary value represents the volume of ratings done by person i	For reviewer k 1—if $NumRating_{k1} >$ Median $(NumRating_{i1})$

(Continued)

Table 1. (*Continued*)

Variable	Definition	Calculation
	at t_1 compared to the median of others' activity	0—if $NumRating_{k1} <$ Median $(NumRating_{i1})$
$SNSize_{ij}$	The number of two-sided friendship bound formed between person i and other reviewers between t_0 and t_j	collected from reviewers' profile
$SNSizeLevel_{ij}$	Is a binary value represents the size of Social Network person i at t_1 are connected to compared to other reviewer	For reviewer k 1—if $SNSize_{k1} >$ Median $(SNSize_{i1})$ 0— if $SNSize_{k1} <$ Median $(SNSize_{i1})$

Table 2. Descriptive statistics

Variable	Mean	Median	SD
$Continouity_{i2}$	0.65	1	0.47
$LowActivityRatio_{i2}$	0.65	0.79	0.46
$AverageDaysBetweenRatings_{i1}$	6.6	4.6	12.1
$ExpectedRatings_{i2}$	31.76	23.12	31.33
$AvgRating_{i1}$	3.82	3.79	0.39
$AvgRatingSidedness_{i1}$	0.5	0.5	0.5
$NumRating_{ij}$	430.5	307	395.1
$NumRatingLevel_{i1}$	0.49	0	0.5
$SNSize_{ij}$	127.6	38	381.1
$SNSizeLevel_{ij}$	0.49	0	0.5

5 Data Analysis

In order to investigate our hypotheses, we need to compare reviewer with different behavior to each other. First we calculated the binary value for three categorical measures of $NumRatingLevel_{i1}$, $SNSizeLevel_{ij}$, and $AvgRatingSidedness_{i1}$ (detail is presented in Table 1). Then we clustered reviewers into four main clusters and labeled them based on the conceptual model (Fig. 1) to Insider, Devotee, Mingler, and Tourist. Each cluster has two sub-cluster for reviewers with positive and negative sidedness. Table 3 includes summary data on contribution measure for all clusters. Overall, we have 620 reviewers in eight clusters. In this study, we do not have any data about the distribution of the dependent variables, neither $Continouity_{i2}$ nor $LowActivityRatio_{i2}$.

The reason is we cluster reviewers in different clusters and the number of users in each cluster is not necessarily enough for the assumption of having a normal sample in each cluster. As we cannot assume normality of their distribution, we cannot use any parametric statistical analysis method. Therefore, we have used two nonparametric statistical tests to compare the distribution of DVs between clusters and evaluate our hypotheses.

Table 3. Measures of interest and descriptive statistics for each cluster

Cluster name	Cluster	$Continouity_{i2}$ (average)		$LowActivityRatio_{i2}$ (average)	
Insider	Cluster1 (Insider −)[a]	0.83	0.8	0.53	0.55
	Cluster3 (Insider +)	0.76		0.59	
Devotee	Cluster5 (Devotee +)	0.73	0.78	0.67	0.61
	Cluster7 (Devotee −)	0.81		0.57	
Mingler	Cluster2 (Mingler +)	0.705	0.61	0.64	0.69
	Cluster8 (Mingler −)	0.47		0.76	
Tourist	Cluster4 (Tourist −)	0.52	0.46	0.73	0.76
	Cluster6 (Tourist +)	0.41		0.79	

[a]Insiders with overall negative sidedness

We used *two-samples Kolmogorov–Smirnov* [38] to compare the distribution of DVs. This test compares the empirical distribution of two independent samples were drawn. The H0 is that both samples are coming from an identical distribution and the calculated P-value are based on asymptotic distributions and represents the confidence interval (CI) under which we can reject H0. This test only compares the distribution and does not include any information about the distribution itself.

KS test has a limitation that can could not solely depend on it. This test is very conservative. The approximations of asymptotic distributions are not to be fully trusted in two situations: small samples and continuous variables [38]. We believe that we still can use the test, but we have double-checked the conservative results with another test, Wilcoxon Rank-sum test.

Wilcoxon rank-sum test compares the distribution of two independent samples, which are unmatched drawn from the same population. The assumptions for this non-parametric test is the continuous of the variable, ordinal response, and independency of two samples. The outcome of the test is a p-value which represents the confidence interval (CI) under which the H0 of same distribution for two samples can be rejected. The test also gives back a probability under which the average of DV for the sample is greater than the second sample [39]. We used two non-parametric tests (KS and Wilcoxon test) and we observed consistent results from both methods. We also made sure that all assumptions of these tests are met in our dataset. It is worth mentioning that as we calculated all independent variables at t_1 and dependent variables between t_0 and t_1 we mitigate the problem of heterogeneity in our analysis.

6   Discussion and Conclusion

Our analysis shows the significance of heterogeneity in product reviewing behavior and predicting its effects on reviewers' ongoing contribution.

The KS and Wilcoxon rank-sum test result (3rd, 5th, and 6th rows in the Table 4) supports our first Hypothesis. It shows that contribution measures for Tourists is coming from a significantly different distribution from Insiders and Devotees (with a high CI). However, this distribution difference between Tourists and Minglers is only significant under CI of 90 %. The Wilcoxon probability also shows that Tourists are more likely to show lower activity compare to Insiders (68 %), Minglers (% 58), and Devotees (% 66)[1]. We conclude that Tourists, who have weak social tie to the community and less interest in the product, are more likely to decrease their contribution over time.

Table 4. Ksmirnov and Wilcoxon run-sum test; comparing reviewer types

Row	First cluster	Second cluster	KSmirnov P value (continuity)	KS p value (low activity)	Wilcoxon run-sum P value (low activity)	Wilcoxon probability
1	Insider	Devotee	1	0.07 ~	0.13	P(Devotee > Insider) = 0.55
2	Insider	Mingler	0.008 **	0.002 **	0.0003***	P(Mingler > Insider) = 0.62
3	Insider	Tourist	0.0***	0.0 ***	0.0 ***	P(Tourist > Insider) = 0.68
4	Devotee	Mingler	0.05 ~	0.03 *	0.02*	P(Devotee > Mingler) = 0.41
5	Devotee	Tourist	0.0 ***	0.0 ***	0.0 ***	P(Devotee > Tourist) = 0.34
6	Mingler	Tourist	0.06 ~	0.02 *	0.01*	P(Mingler > Tourist) = 0.42

$P < 0.001$***, $P < 0.01$ **, $P < 0.05$ *, $P < 0.1$ ~

The second hypothesis focuses on the centrality of the consumption activity for reviewers. We believe that consumption activity, itself, can drive the contribution of reviewers, as their interest in the product is not a function of time or social feedback. The result of our analysis (2nd and 5th rows in Table 4), supports this argument for reviewers with strong social ties at CI of 99.99 %. The same result also supports the hypothesis for reviewers with weak social tie at 99 %. However, the result does not fully support H3 (1st, 2nd, and 3rd rows in the Table 4). We expected that Insiders, who have a high level of consumption activity and strong ties to the society, maintain significantly higher contribution than all other members do. We showed that Insiders' ongoing contribution is significantly different from Minglers and Tourists (with the probability of 0.62 and 0.68). Yet their difference with Devotees is not significant. This result is in line with the result of H2, which emphasizes the consumption activity as the driver of ongoing contribution. However, we have built our hypothesis on the previous literature [3, 7]

---

[1] 100*(1–0.42) for Minglers and 100*(1–0.34) For Devotees.

based on which we expected to observe the effect of popularity as the important driver of reviewing volume. We also can see the same effect in the analysis for H4 (1st and 6th rows in the Table 4). We cannot reject the null hypothesis that the contribution of Insiders and Devotees are coming from different distributions. However, we observed that Minglers and Tourists are performing different at the CI level of 90 %.

It is more complicated to examine H5. We intend to study the effect of sidedness in changing the ongoing contribution. To do so, first we ran both KS and Wilcoxon tests, for each cluster at a time, and then compared the contribution level of that cluster to the rest of our reviewers (Table 5). We have observed that the contribution of Insiders with negative sidedness (1st row in the Table 5) and tourists with positive sidedness are coming from different distributions with very high CI (99.99 %). This effect is not this strong for any other clusters. Therefore, we concluded that the effect of sidedness on ongoing contribution is significant for extreme reviewers.

Table 5. KS and Wilcoxon run-sum test; comparing each cluster with the whole sample

Row	Description	Cluster	# of reviewers	KS test P value	Mean (0–1)	KS p value-low activity	Wilcoxon P value	Wilcoxon probability
1	Tourist with positive sidedness	6	112	0.00 ***	0.41	0.0***	0.000 ***	0.33
2	Insider with negative sidedness	1	108	0.001**	0.83	0.0 ***	0.0 ***	0.6
3	Tourist with negative sidedness	4	85	0.068 ~	0.52	0.04 *	0.02 *	0.42
4	Insider with positive sidedness	3	85	0.194	0.76	0.002**	0.005 **	0.59
5	Devotee with negative sidedness	7	72	0.030 *	0.81	0.02 *	0.018 *	0.58
6	Mingler with positive sidedness	2	68	0.991	0.7	0.76	0.57	0.521
7	Mingler with negative sidedness	8	46	0.069 ~	0.47	0.035 *	0.011 *	0.39
8	Devotee with positive sidedness	5	45	0.924	0.73	0.59	0.75	0.51

$P < 0.001$***, $P < 0.01$ **, $P < 0.05$ *, $P < 0.1$ ~

Table 6. Summary of analysis results

Hypothesis	Status
Hypothesis 1: In an online community of consumption, members with weaker social ties and weaker consumption activity (Tourists) are more likely to have a lower ongoing contribution comparing to other members	Supported
Hypothesis 2: In an online community of consumption, for members with the similar level of social ties, level of ongoing contribution will be driven by the levels of their consumption activity	Supported
Hypothesis 3: In an online community of consumption, members with stronger social ties and higher consumption activity (Insiders) are more likely to have a higher ongoing contribution comparing to other reviewers	Partially supported
Hypothesis 4: In an online community of consumption, for members with similar consumption activity levels, level of ongoing contribution will be driven by the strength of their social ties	Partially supported
Hypothesis 5: In an online community of consumption, members with negative overall sidedness are more likely to have a higher ongoing contribution comparing to other reviewers	Partially supported (only in extreme situations)

We know that insiders have high consumption activity. We showed that members, who have negative sidedness, have more chance to get the negatively biased social feedback. Therefore, they may have more incentives to continue their contribution. This is completely in line with Goes et al. [3] and the popularity effect. On the other extreme, Tourists have less chance to get some social feedback as they do not have many audiences and did not expose many of their opinions. However, Tourists who have positive sidedness are less likely to get any social feedback, which we know is negatively biased [33] and it is more likely for them to leave the community sooner. In addition, Tourists with negative sidedness might have extra motivation to write about their negative consumption experience and consequently stay more than other Tourists (second row in the Table 5).

7   Conclusion and Implication

We have used the theory of e-tribulized marketing and different member types in an online community of consumption to study an eWOM community of book reviews. We confirmed some driving factors of ongoing reviewing contribution. We proposed a three-dimensional conceptual model to cluster different reviewing behavior patterns. We showed that the heterogeneity in reviewing pattern, especially in leaving the platform could be predicted by these three dimensions: *Strength of social ties, level of consumption activity*, and *reviewer's sidedness*. We showed that the effect of sidedness on contribution prediction is stronger for reviewers with extreme reviewing behavior.

We also concluded that consumption activity has more predictive information about the contribution compare to the social tie and sidedness. Our results are aligned with both Kozitents' work [2] and the popularity effect of eWOM [3] (Table 6).

This research, like any other research has limitation. The main limitation is the static nature of the analysis, which does not cover the time-related dynamic of changing behavior in reviewers. Moreover, the effect of product type was controlled in this research and we will continue this research removing that limitation in the next step.

This result can have a high implication for online review hosting platforms. In order to maintain a sustainable online community, these platforms must have some tailored policies, rewards, and other mechanisms to attract and maintain their members. We believe that using the result of this research, these mechanisms can be tailored for individual members based on their historical reviewing behavior to maximize the likelihood of their continuous contribution.

References

1. Iriberri, A., Leroy, G.: A life-cycle perspective on online community success. ACM Comput. Sur. (CSUR) **41**(2), 11 (2009)
2. Kozinets, R.V.: E-tribalized marketing?: The strategic implications of virtual communities of consumption. Eur. Manag. J. **17**(3), 252–264 (1999)
3. Goes, P., Lin, M., Yeung, C.: "Popularity effect" in user-generated content: evidence from online product reviews. Inf. Sys. Res. 1–17 (2014)
4. Cheung, C.M.K., Thadani, D.R.: The impact of electronic word-of-mouth communication: a literature analysis and integrative model. Decis. Support Syst. **54**(1), 461–470 (2012)
5. Mudambi, S.M., Schuff, D.: What makes a helpful online review? A study of customer reviews on Amazon.com. MIS Q. **34**(1), 185–200 (2010)
6. Wei, X., Chen, W., Zhu, K.: Motivating user contributions in online knowledge communities: virtual rewards and reputation. In: 48th Hawaii International Conference on System Sciences (2015)
7. Samiei, P., Tripathi, A.K.: Effect of social networks on online reviews. In: 47th Hawaii International Conference in System Sciences (HICSS), Hawaii (2014)
8. Connors, L., Mudambi, S.M., Scuff, D.: Is it the review or the reviewer? A multi-method approach to determine the antecedents of online review helpfulness. In: 44th Hawaii International Conference on System Sciences (HICSS) (2011)
9. Wasko, M.M., Faraj, S.: Why should I share? Examining social capital and knowledge contribution in electronic networks of practice. MIS Q. **29**(1), 35–57 (2005)
10. Alexandrov, A., Lilly, B., Babakus, E.: The effects of social-and self-motives on the intentions to share positive and negative word of mouth. J. Acad. Mark. Sci. **41**(5), 531–546 (2013)
11. Brooks, R.C.: "Word-of-Mouth" advertising in selling new products. J. Mark. **22**(2), 154–161 (1957)
12. Moe, W.W., Schweidel, D.A.: Online product opinions: incidence, evaluation, and evolution. Mark. Sci. **31**(3), 372–386 (2012)
13. Munzel, A., Kunz, W.H.: Creators, multipliers, and lurkers: who contributes and who benefits at online review sites. J. Serv. Manage. **25**(1), 49–74 (2014)

14. Cheung, C.M., Lee, M.K.: What drives consumers to spread electronic word of mouth in online consumer-opinion platforms. Decis. Support Syst. **53**(1), 218–225 (2012)
15. Hennig-Thurau, T., Gwinner, K.P., Walsh, G., Gremler, D.D.: Electronic word-of-mouth via consumer-opinion platforms: what motivates consumers to articulate themselves on the Internet? J. Interact. Mark. **18**(1), 38–52 (2004)
16. Hu, N., Pavlou, P.A., Zhang, J.: Can online reviews reveal a product's true quality? Empirical findings and analytical modelling of online word-of-mouth communication. In: Proceeding of 7th ACM Conference, Electronic Commerce, New York (2006)
17. Li, X., Hitt, L.M.: Self-selection and information role of online product reviews. Inf. Syst. Res. **19**(4), 456–474 (2008)
18. Zhu, F., Zhang, X.: Impact of online consumer reviews on sales: the moderating role of product and consumer characteristics. J. Mark. **74**(2), 133–148 (2010)
19. Zhang, X., Dellarocas, C.: The lord of the ratings: is a movie's fate is influenced by reviews? In: ICIS 2006 Proceedings (2006)
20. Shen, W.: Competing for attention in online reviews. In: AMCIS 2009 Doctoral Consortium (2009)
21. Huang, J., Cheng, X.Q., Shen, H.W., Zhou, T., Jin, X.: Exploring social influence via posterior effect of word-of-mouth recommendations. In: Fifth ACM International Conference on Web Search and Data Mining (WSDM 2012) (2012)
22. Montazemi, A.R., Saremi, H.Q.: The effectiveness of electronic word of mouth on consumers' perceptions of adopting products/services (2014)
23. King, R.A., Racherla, P., Bush, V.D.: What we know and don't know about online word-of-mouth: a review and synthesis of the literature. J. Interact. Mark. **28**(3), 167–183 (2014)
24. Ridings, C., Gefen, D., Arinze, B.: Psychological barriers: Lurker and poster motivation and behavior in online communities. Commun. Assoc. Inf. Syst. **18**(1), 16 (2006)
25. Takahashi, M., Fujimoto, M., Yamasaki, N.: The active Lurker: a new viewpoint for evaluating the influence of an in-house online community. ACM SIGGROUP Bull. **23**(3), 29–33 (2002)
26. Li, H., Lai, V.: The interpersonal relationship perspective on virtual community participation. In: ICIS 2007 Proceedings (2007)
27. Hartmann, B.J., Wiertz, C., Arnould, E.J.: Exploring consumptive moments of value-creating practice in online community. Psychol. Mark. **32**(3), 319–340 (2015)
28. Preece, J., Shneiderman, B.: The reader-to-leader framework: motivating technology-mediated social participation. AIS Trans. Hum. Comput. Interact. **1**(1), 13–32 (2009)
29. Kozinets, R.V.: Utopian enterprise: articulating the meanings of Star Trek's culture of consumption. J. Consum. Res. **28**(1), 67–88 (2001)
30. Butler, B.S.: Membership size, communication activity, and sustainability: a resource-based model of online social structures. Inf. Syst. Res. **12**(4), 346–362 (2001)
31. Wang, Y.C., Kraut, R., Levine, J.M.: To stay or leave?: The relationship of emotional and informational support to commitment in online health support groups. In: ACM 2012 Conference on Computer Supported Cooperative Work Proceeding (2012)
32. Nov, O., Naaman, M., Ye, C.: Analysis of participation in an online photo-sharing community: a multidimensional perspective. J. Am. Soc. Inform. Sci. Technol. **61**(3), 555–566 (2010)
33. Schlosser, A.E.: Posting versus lurking: communicating in a multiple audience context. J. Consum. Res. **32**(2), 260–265 (2005)
34. Samiei, P., Tripathi, A.K.: Exploring reviewers' contributions to online review platforms. In: 23rd Workshop on Information Technologies and Systems, Milan, Italy (2013)

35. Xu, Y.C., Zhang, C., Xue, L.: Measuring product susceptibility in online product review social network. Decis. Support Syst. (2013)
36. Samiei, P.: Understanding online reviewing behavior, product evaluation. In: 24th Workshop on Information Technologies and Systems (WITS), Auckland, New Zealand (2014)
37. Chen, H.-N., Huang, C.-Y.: An investigation into online reviewers' behavior. Eur. J. Mark. **47**(10), 1758–1773 (2013)
38. Stata.com: Ksmirnov- Kolmogorov –Smirnov equality-of-distributions test. In: Stata manual (2015)
39. Huang, D.S., Zhao, Z., Bevilacqua, V., Figueroa, J.C.: Advanced intelligent computing theories and applications. In: 6th International Conference on Intelligent Computing, ICIC 2010, Changsha, China (2010)

The Role of Web and E-Commerce
in Poverty Reduction: A Framework
Based on Ecological Systems Theory

Dong-Heon Kwak[1](✉) and Hemant Jain[2]

[1] Kent State University, Kent, USA
dkwak@kent.edu
[2] University of Wisconsin-Milwaukee, Milwaukee, USA
jain@kent.edu

Abstract. Web and eCommerce enabled by easy access to Internet on mobile devices have a great potential to reduce poverty by improving access to education, health, government, financial and other services, and by providing access to potential global markets for the products and services they can offer. However, the role of Web/eCommerce in poverty reduction has not been well studied in IS research; specifically no theoretical framework for such studies exist. The purpose of this study is to develop a theoretical framework that can help identify causes of poverty and help examine how Web/eCommerce can intervene to reduce poverty. Our framework is based on Bronfenbrenner's ecological systems theory. We apply the resulting framework to rural farming families.

Keywords: Poverty reduction · Ecological systems theory · Web · eCommerce · ICT

1 Introduction

Poverty has been an endless concern in the human history. According to Bruton [8], "there remain stubborn levels of poverty among the bottom sixth of the world's population, with an estimated one billion people continuing to live on less than $1 a day on average" ([8], p. 6). The complex, multidimensional, ubiquitous nature of poverty has led researchers in various disciplines (e.g., economics, sociology, anthropology, psychology, education) to identify various strategies to reduce poverty. Recently, there have been efforts by management researchers to address the poverty issue. In particular, businesses could play an important role in reducing poverty while making profits if they adapted their business models to serve population at the Bottom of Pyramid [33]. Web and eCommerce enabled by easy access to Internet through mobile devices can allow businesses to reach remarkable new markets which consist of billions of people at lower end of income spectrum. Additionally, it can open up new markets for products and services provided by population at the Bottom of Pyramid. This can help alleviate the desperate poverty [30].

Information and communication technologies (ICTs), specifically Web access and eCommerce enabled by mobile devices, can play a significant role as enabling technology for reducing poverty (for ease of reference we use the term ICTs to refer to

© Springer International Publishing Switzerland 2016
V. Sugumaran et al. (Eds.): WEB 2015, LNBIP 258, pp. 143–154, 2016.
DOI: 10.1007/978-3-319-45408-5_12

WEB access and eCommerce enabled by mobile devices). Research on the effects of ICTs on poverty reduction has been growing [1]. There have been many ICT-based development projects initiated by international aid institutions (e.g., UN, World Bank), nonprofit organizations, and educational institutions. Although some scholars believe that ICTs contribute to wider economic divergence between developed and developing countries [4], others regard it as critical means for helping the poor and reducing poverty. ICTs have potential to reduce poverty by improving poor people's access to education, health, government, financial services, and relevant information [9, 15, 16, 28, 29, 33, 38]. Web, for example, can help small farmers in rural areas by connecting them to market or providing easy access to relevant and accurate agricultural information [9].

We argue that use of ICTs for poverty reduction needs to be theoretically studied from the academic perspective; and based on this, appropriate systems needs to be designed, developed, implemented, used, and maintained. However, ICTs and poverty reduction have been rarely studied in Information Systems (IS) discipline. In addition, there is a lack of theoretical framework that can cover large number of reasons of poverty. Given the importance of the research on the relationship between ICTs and poverty reduction and lack of previous research in IS, the purpose of this study is to develop a theoretical framework that identify causes of poverty and examine how ICTs can intervene to reduce poverty. We use Bronfenbrenner's ecological systems theory (EST) [6] as a basis for developing our framework.

The paper is organized as follows. In Sect. 2, we present review of literature on ICTs and poverty reduction. In Sect. 3, EST as an overarching theory is reviewed and theoretical framework is presented. In Sect. 4, we use the EST framework to study causes of poverty in farming family and examine the role of ICTs in helping reduce poverty. Future research directions with possible research questions are provided in Sect. 5.

2 ICT and Poverty Reduction

ICTs are defined as "technologies that can process different kinds of information and facilitate different forms of communications among human agents, among humans and information systems, and among information systems" ([20] cited in [11], p. 6). This definition is consistent with "*tool* view of technology" suggested by Orlikowski and Iacono [31]. Since our focus is on investigating how ICTs can alleviate poverty, the *tool* view of technology is the core conceptualization of ICTs used for poverty reduction. Adeya [1] reviews studies on how old ICTs (e.g., radio and telephone) and new ICTs (e.g. Internet, cell phone, and computer based technologies) contribute to poverty reduction. Kenny [27] argues that a variety of old and new ICTs can alleviate poverty by providing analysis of costs and benefits of ICTs. Radio has been regarded as powerful means for spreading information and educating poor people in both urban and rural areas [1]. Studies are still being conducted on the benefits of radio in the information age (e.g., [27]). Examples of new ICTs include computerized milk collection centers that support small poor dairy farmers [9] and telemedicine to reduce the cost and hardship of long distance travel for the poor in rural area [29].

It should be noted that there have been both skeptic and optimistic views on whether ICTs can help the poor. Two opposite perspectives on the impact of ICTs on poverty are briefly discussed here. The skeptic position is that ICTs are no more than tool and they have no influence on the poverty reduction. Mansell and Wehn [28] warn that if developing countries initiate ICT implementation strategies that imitate the one person – one telephone – one Internet access point model dominated in developed countries, frustration will be widespread instead of helping the poor. Furthermore, Brown [7] states that ICTs are merely tools, arguing that no single tool can resolve global, complex, multidimensional problems such as poverty. Chowdhury [11] mentions some skeptic view of ICTs such that "ICTs do not have any more to do with poverty and food security in the developing countries than rain dances have to do with rain." Additionally, he argues that poor people cannot afford high-speed Internet access. Braga [4] argues that ICTs result in a wider economic gap between developed and developing countries.

On the other hand, many scholars have pointed out the positive effects of ICTs on the poor. Prahalad [33] states that "there are now a large number of examples of organizing the poor to ensure that they have the benefits of information-as in the case of farmers using cell phones to check weather and price information before they sell to farming cooperatives or working with large firms such as ITC or Nestle" (p. 23). Heeks and Bhatnagar [22] state that ICTs for poverty reduction primarily play a role of communications technologies rather than of information-processing or production technologies, suggesting that access to information is a priority in helping the poor. In addition, ICTs can help small and medium-scale enterprises by reducing transaction costs and improving communications with markets and within the supply chain [16]. Community information centers can help the poor in rural areas by providing relevant information [37]. Moreover, ICTs are seen as having potential to effectively combat pandemic such as HIV/AIDS [1, 13].

Since poverty is a complex, multidimensional, ubiquitous phenomenon, it is not possible for researchers to examine all causes of poverty and all types of poor people. Thus, research to date has focused on the relationship between ICTs and poverty reduction by selecting specific target (e.g. women in developing countries) or topic (e.g., market access, education, health).

3   Theoretical Development: Ecological Systems Theory

This study employs Bronfenbrenner's ecological systems theory (EST) to examine causes of poverty and impact of ICTs on poverty reduction. Bronfenbrenner developed EST in an effort to define and identify human development. According to Bronfenbrenner [6]:

The ecology of human development is the scientific study of the progressive, mutual accommodation throughout the life course between an active, growing human being and the changing properties of the immediate settings in which the developing person lives. [This] process is affected by the relations between these settings and by the larger contexts in which the settings are embedded (p. 188).

EST suggests that human development occurs through continuous, reciprocal interactions between human beings and the individuals, objects, and symbols in the environment [5]. The environment is comprised of five layers of systems which interact in complex ways and has bi-directional influences within and among systems [5]. The five systems are microsystem, mesosystem, exosystem, macrosystem, and chronosystem. The microsystem refers to the environment which immediately affects a person. It includes parents, peers, home, and others. The mesosystem is comprised of interactions among two or more immediate environments. Examples of mesosystems are relations between the child's teacher or peer group and the parents. Experiences in a microsystem, such as parent-child interactions in the home may affect activities and interactions in another, such as the interaction with teacher, or vice versa. Thus, possible linkages among microsystems are countless. Exosystem indirectly influence a person by directly influencing microsystem and mesosystem. The school and the school board are examples of mesosystem in child development; events that occur in the decision of school board can have consequences for the teacher in the school. Macrosystem refers to the dominant social ideologies and cultural values that partially determine the social structure and activities. The chronosystem emphasizes the effect of time on all systems and all developmental processes, including consistency or change over the life course [26]. The examples of chronosystem are parental divorce and historical events. According to Eamon [17], scholars have shown the impacts of income loss from historical events such as the Great Depression [18] and the 1980s Midwest farm crisis [12] on children's socio emotional development. Figure 1 illustrates examples of each system in EST.

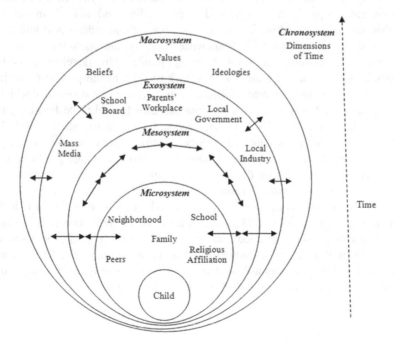

Fig. 1. Bronfenbrenner's ecological systems theory

EST has been applied in various contexts such as hard-line treatment of asylum seekers [32] and demand for mental health service by Asian American [36]. This theory can be extended to model the development of an organization as well [25]. EST provides a good framework to examine the relationship between ICTs and poverty. Since ICTs can impact all the five systems in the EST, this theory is well suited to examine the overall impact of ICTs on poverty. Research on poverty has generally focused on micro or single topic. EST provides broad map which can include micro to macro aspects of causes of poverty. In the next section we use this framework to study the causes of poverty of farming family who live in rural areas in developing countries. We then examine how ICTs can impact various systems related to farming family and in turn impact poverty.

4 Environment of Poor Rural Farming Families in Developing Countries and Impact of ICT

To identity the role of ICTs in alleviating poverty, this study focuses on farming families living in rural areas in developing countries (cf. [38]). The World Bank's flagship annual report on development, places the productivity of small farmers at the core of a global agenda to alleviate poverty [14]. While three-quarters of the developing world's 1.4 billion extremely poor people live in rural areas, only 4 percent of official development support goes to agriculture in developing countries [14, 24].

There are many possible causes of poverty of farming families; in this paper we focus on the causes which influence their income and cost. In other words, we focus on the reasons that prevent them from increasing their income and make them spend more money than required. The strategies to increase income and reduce cost will result in poverty reduction. Based on EST framework microsystem includes factors which directly influence poverty of rural farming families in developing countries. Microsystem includes lack of market access, lack of skill, health problem, and lack of relevant information. Mesosystem is interactions between microsystems which also directly impact poor farmers. Exosystem does not directly influence farming families but it has indirect relationships with mesosystem and microsystem that influence farmers. Lack of institutional help and regional enterprise are examples of exosystem. Macrosystem includes country status, infrastructure, and language problem that has impact on the whole society. A detailed list of causes of poverty of farming family based on EST model is shown in Table 1.

Given the above causes of poverty based on EST, we now examine how ICTs can intervene at various systems in EST model to help poverty reduction in the context of rural farming family. We examined and analyzed existing literature on various ICT applications in rural areas and classified those applications using the EST framework described above. This provides a theoretical basis and deeper understanding into how ICT applications impact poverty reduction efforts (Table 2).

Table 1. Causes of poverty of rural farming families in developing countries

System	Causes of poverty	Explanation
Microsystem	Market access problem	Lack of market access prevents farmers from having opportunities to sell their products
	Health problem - Disease - Insufficient nutrition - Sanitation	Health problem implies loss of labor force and increased medical cost
	Educational Problem - Lack of skill - Illiteracy	Farmers who do not have sufficient agricultural skills can have less productivity. Illiteracy prevents farmers from learning new knowledge
	Lack of information	Lack of relevant information (e.g. crop disease, weather forecast) can negatively influence agricultural productivity
Exosystem	Institutional help - International aid agency (UN, World Bank) - Donor community (Nonprofit organizations)	There are many institutions for development. They provide educational and health information. Physical supports are provided as well. Their assistance can partly solve the problems in microsystem
	Lack of competency of regional enterprises	Small regional enterprises can contribute to the community. Boosting small enterprises results in increasing market opportunity and other information
	Community problem	It is difficult to access relevant information and have better market access in less competent community
Macrosystem	Country status	Compared to developed countries, developing countries are at a disadvantage because they have less developed infrastructure and many sources of information are from developed countries
	National infrastructure - Restricted accommodation - Lack of communication channels - Lack of computer systems	National infrastructure problems influence exosystems and microsystems. For example, extremely poor infrastructure prevents institutional help by making it hard for them to access the country
	Illiteracy rates, Ignorance	National illiteracy rate and ignorance influences many causes of poverty described above
	Geographic isolation	Geographically isolated regions are at relative disadvantage to access information and to get help from other institutions

(Continued)

Table 1. (*Continued*)

System	Causes of poverty	Explanation
	Language problems - Multiple languages -Source of information (English/developed country)	It could be hard to provide information and educational services effectively in regions with multiple languages. Currently most information and ICT systems are produced in English by developed countries. This could prohibit information access and increase learning problems
Chronosy-stem	Past experience - War - Past illness - Environmental catastrophe	Past experiences can impacts current status of people. For example, countries that experienced severe war can be still poor

Table 2. Intervention of ICTs and ecosystems for intervention

ICTs	Description	Ecosystems for intervention
Computerized milk collection centers in India [9]	Dairy Information System Kiosk software developed by Indian Institute of Management, Ahmedabad offers useful information to farmers via a database including complete histories of all milk cattle owned by members of the cooperative and a dairy portal connected to the Internet. Dairy farmers who incorporate the computerized system benefit from a more efficient cooperative system	**Microsystem** (Information access)
Auxiliary nurse midwives' health delivery project in India [9]	Personal Digital Assistants (PDAs) allow auxiliary nurse midwives (ANMs) participating in the Indian Healthcare Delivery project for alleviating redundant paperwork and data entry, freeing up time for healthcare delivery to poor people. PDAs facilitate data collection and transmission, saving up to 40 percent of ANMs' work time. This facilitates access to basic services	**Exosystem** (Institution)

(*Continued*)

Table 2. (*Continued*)

ICTs	Description	Ecosystems for intervention
Crop disease forecasting system in India [34]	This is a part of development project supported by Asia Pacific Development Information Program (APDIP). The project utilizes existing kiosks and mobile networks to make information available. Local community radio station is also incorporated in the system to benefit farmers. The system distributes crop advisory and provides disease forecasting services to farmers in a resource-limited environment in India. This project aims at increasing the accuracy of forecasts to farmers resulting in increased productivity and income	**Microsystem** (Information access) **Exosystem** (Community)
Web-based e-crop management in China [39]	This is a part of APDIP project to develop an "e-Farm" system which offers dynamic web-based crop information and simulation services. This results in a 30 % reduction in nitrogen application in rice production, representing substantial savings in labor and fertilizer costs	**Microsystem** (Information access) **Exosystem** (Local government and educational institute)
Mobile telemedicine system in Indonesia [35]	This is a part of APDIP project to develop an ICT based mobile telemedicine system. The system is for patients who live far away from a hospital and local hospitals have limited human resources. The system can also overcome the geographic problems	**Microsystem** (Health), **Exosystem** (Local hospital) **Macrosystem** (Geographic Isolation)
English-Nepali Translator in Nepal [3]	This is a part of APDIP project to develop a web-based engine that provides the translation from English to Nepali. The program is for Nepali speaking Internet users and other institutions. This can make additional information available to people and institutions	**Microsystem** (Information access) **Macrosystem** (Language Problem)
Flower farmer in India [23]	A flower farmer in India has reduced his workload while more than doubling his monthly income because of better market price information through his mobile phone	**Microsystem** (Information access)

(*Continued*)

Table 2. (*Continued*)

ICTs	Description	Ecosystems for intervention
Smartphone [2]	Smartphones in Uganda are helping thousands of poor farmers to track new farming technologies, treatment for their animal, weather patterns, market prices and best bargains	**Microsystem** (Information access)
Village phone in Uganda [21]	Village phone boosted local economy because people in the community can communicate easily and find small business opportunities via information exchange	**Exosystem** (Community)

5 Discussion, Conclusion and Future Research

This study is one of the few studies to investigate the role of ICTs in poverty reduction in IS discipline. This study incorporates EST which can explain individual and societal structure simultaneously. EST proved to be useful and comprehensive theoretical framework in examining the complex nature of poverty. Drawing upon EST, this study investigated how ICTs can reduce poverty. The EST theoretical framework helps classify a development project into one or more systems of the framework and thus evaluate how the project can impact poverty reduction goal. We have reviewed a number projects done in various countries and classified them based on the framework. This shows the viability and usefulness of the framework.

Below we provide future research directions and research questions that can be examined. We focused on poverty of farming families because of the importance of investigating rural farmers in developing countries. However, there are other types of poor people living in different areas. ICTs can help them in various ways such as education and health information. Future researchers can select potential target poor population based on subjects (e.g., child, women, and small- and medium-sized enterprise), area (e.g., rural or urban), country type (e.g., developing or developed), and theme (education, health, gender, or equality). With respect to these following research questions can be examined.

Research Question 1: *In examining the role of ICTs in poverty reduction, what population of poor people should be studied and why?*

This study provided some evidences of positive effects of ICT on poverty reduction. Future researchers can incorporate various methods (e.g., case study, longitudinal study, and action research) to prove the role of ICTs in reducing poverty. In addition, future researchers can develop systems for poverty reduction and validate it. Systems dynamics models (e.g. [10]) can be one of the methods for validation.

Research Question 2: *What methodological issue exists in examining the effects of ICTs on poverty reduction?*

Many institutions such as international aid institutions, charity organizations, research institutions, universities, and other for profit businesses have initiated projects to develop ICTs for poverty reduction. Given the assumption that ICTs can reduce poverty, the diffusion of ICTs is a critical issue for achieving successful poverty reduction. Therefore, future researchers need to study issues related to ICT diffusion in bottom of the pyramid market:

Research question 3: *What diffusion issues exist in ICTs and poverty reduction?*

Gill and Bhattacherjee [19] suggest informing challenge in IS discipline. The scholars consider how the IS discipline is serving three clients: the practitioner client, the student client, and clients from other disciplines. Since poverty is a ubiquitous problem, various types of practitioners are struggling to address it. International institutions (e.g., UN, World Bank), charity organizations (e.g., World Vision), and other businesses are among them. Many other academic disciplines are doing research on poverty. To effectively inform the role of ICTs on poverty reduction, we should consider our clients as suggested by Gill and Bhattacherjee.

Research question 4: *What clients should we focus in conducting research on ICTs and poverty reduction? How can we inform them to achieve practical poverty reduction?*

Given some qualitative evidences that ICTs can reduce poverty, this study provides future researcher various research questions that are worth examining. Addressing a complex problem such as poverty requires combination of efforts from different academic disciplines and practitioners. We hope this study will motivate future poverty reduction research.

References

1. Adeya, N.C.: ICTs and Poverty: A Literature Review. International Development Research Centre (2002)
2. Jazeera, Al.: English Smartphones come to Ugandan farmers' aid [video file] (2013). https://www.youtube.com/watch?v=pHWcFLyB25w. Accessed 15 Aug 2015
3. Bista, S.K.: Dobhase: English-Nepali translator. Asia-Pacific Development Information Programme (2006). http://www.apdip.net/resources/case/rnd32/view. Accessed 11 May 2011
4. Braga, C.: Inclusion or Exclusion? UNESCO Courier (1998). http://unesdoc.unesco.org/images/0011/001142/114252e.pdf. Accessed 20 May 2015
5. Bronfenbrenner, U.: Developmental ecology through space and time: a future perspective. In: Moen, P., Elder Jr., G.H., Luscher, K. (eds.) Examining Lives in Context: Perspectives on the Ecology of Human Development, pp. 619–647. American Psychological Association, Washington, D.C. (1995)
6. Bronfenbrenner, U.: Recent advances in research on human development. In: Silbereisen, R.K., Eyferth, K., Rudinger, G. (eds.) Development as Action in Context: Problem Behavior and Normal Youth Development, pp. 287–309. Springer, New York (1986)
7. Brown, M.M.: Can ICTs address the needs of the poor? A commentary from UNDP (2001). http://www.undp.org/dpa/choices/2001/june/j4e.pdf. Accessed 26 July 2011

8. Bruton, G.D.: Business and the world's poorest billion–the need for an expanded examination by management scholars. Acad. Manage. Perspect. **24**, 6–10 (2010)
9. Cecchini, S., Scott, C.: Can information and communications technology applications contribute to poverty reduction? Lessons from rural India. Inf. Technol. Dev. **10**, 73–84 (2003)
10. Choi, J., Nazareth, D.L., Jain, H.K.: Implementing service-oriented architecture in organizations. J. Manag. Inf. Syst. **26**, 253–286 (2010)
11. Chowdhury, N.: Information and Communications Technologies and IFPRI's Mandate: A Conceptual Framework. International Food Policy Research Institute, Washington, D.C., pp. 3–33 (2000)
12. Conger, R.D., Conger, K.J., Elder, G.H., Lorenz, F.O., Simons, R.L., Whitebeck, L.B.: Family economic stress and adjustment for early adolescent girls. Dev. Psychol. **29**, 206–219 (1993)
13. Driscoll, L.: HIV/AIDS and Information and Communication Technologies. International Development Research Centre (2001)
14. Dugger, C.W.: World bank report puts agriculture at core of antipoverty effort. The New York Times. (2007). http://query.nytimes.com/gst/fullpage.html?res=9C04EED6153EF933A15753C1A9619C8B63. Accessed 10 Apr 2013
15. Duncombe, R.A., Boateng, R.: Mobile phones and financial services in developing countries: a review of concepts, methods, issues, evidence and future research directions. Third World Q. **30**, 1237–1258 (2009)
16. Duncombe, R.A., Heeks, R.B.: Information & Communication Technologies (ICTs), Poverty Reduction and Micro, Small & Medium-Scale Enterprises (MSMEs): A Framework for Understanding ICT Applications for MSMEs in Developing Countries. United Nations Industrial Development Organization, Vienna (2005)
17. Eamon, M.K.: The effects of poverty on children's socioemotional development: an ecological systems analysis. Soc. Work **46**, 256–266 (2001)
18. Elder, G.H., Caspi, A.: Economic stress in lives: developmental perspectives. J. Soc. Issues **44**, 25–45 (1988)
19. Gill, G., Bhattacherjee, A.: Whom are we informing? Issues and recommendations for MIS research from an informing sciences perspective. MIS Q. **33**, 217–235 (2009)
20. Hamelink, C.J.: New Information and Communication Technologies: Social Development and Cultural Change. Discussion Paper, vol. 86. UNRISD, New York (1997)
21. Hansen, H.: Uganda village phone field research 2004 [video file] (2015). https://www.youtube.com/watch?v=Ec1VAIlgRbI. Accessed on 15 Aug 2015
22. Heeks, R., Bhatnagar, S.: Understanding success and failure in information age reform. In: Reinventing Government in the Information Age: International Practice in IT-Enabled Public Sector Reform, pp. 49–75 (1999)
23. IDRC: Cell phones can help alleviate poverty in developing countries [video file] (2009). http://www.youtube.com/watch?v=8ARrLWWDjb4. Accessed 6 Mar 2014
24. IFAD: Rural poverty report (2011). http://www.ifad.org/rpr2011/index.htm. Accessed 8 May 2012
25. Johnson, G.M.: Internet use and child development: the techno-microsystem. Aust. J. Educ. Dev. Psychol. **10**, 32–43 (2008)
26. Johnson, G.M.: Internet use and child development: validation of the ecological techno-subsystem. Educ. Technol. Soc. **13**, 176–185 (2010)
27. Kenny, C.: The costs and benefits of ICTs for direct poverty alleviation. The World Bank, Washington, D.C. (2002)

28. Mansell, R., Wehn, U.: Knowledge Societies: Information Technology for Sustainable Development. Oxford University Press, New York (1998). United Nations Commission on Science and Technology for Development. (eds.) Oxford, Published for and on Behalf of the United Nations

29. Miscione, G.: Telemedicine in the upper Amazon: interplay with local health care practices. MIS Q. **31**, 403–425 (2007)

30. Moyo, D.: Dead Aid: Why Aid is not Working and How there is a Better Way for Africa. Farrar Straus, and Giroux, New York (2009)

31. Orlikowski, W.J., Iacono, C.S.: Desperately seeking the "IT" in IT research-a call to theorizing the IT artifact. Inf. Syst. Res. **12**, 121–134 (2001)

32. Pedersen, A., Kenny, M.A., Briskman, L., Hoffman, S.: Working with Wasim: a convergence of community. Aust. Commun. Psychol. **20**, 57–72 (2008)

33. Prahalad, C.K.: The Fortune at the Bottom of the Pyramid: Eradicating Poverty Through Profits. Wharton School, Upper Saddle River (2010)

34. Ramamritham, K.: Crop disease forecasting system and expert crop-advisory to farmers over information kiosk networks in India. Asia-Pacific Development Information Programme (2006). http://www.apdip.net/resources/case/rnd46/view. Accessed 10 May 2011

35. Soegijoko, S.: Development of ICT-based mobile telemedicine system with multi-communication links for urban and rural areas in Indonesia. Asia-Pacific Development Information Programme (2006). http://www.apdip.net/resources/case/rnd55/view. Accessed 10 May 2011

36. Takayama, J.R.: Ecological systems theory of Asian American mental health service seeking. master's thesis, Pacific University (2010)

37. Ulrich, P.: Poverty reduction through access to information and communication technologies in rural area. Electron. J. Inf. Syst. Developing Countries **16**, 1–38 (2004)

38. Venkatesh, V., Sykes, T.A.: Digital divide initiatives success in developing countries: a longitudinal field study in a village in India. Inf. Syst. Res. **24**, 239–260 (2013)

39. Yang, J.: Web-based e-crop management, pacific development information programme (2006). http://www.apdip.net/resources/case/rnd45/view. Accessed 10 May 2011

Using Text Mining Analytics to Understand IT Internal Control Weaknesses

Peiqin Zhang[1(✉)], Lucian L. Visinescu[1], Kexin Zhao[2], and Ram L. Kumar[2]

[1] Department of Computer Information Systems and Quantitative Methods,
McCoy College of Business, Texas State University, 601 University Drive,
San Marcos, TX 78666, USA
{p_z13,llv19}@txstate.edu

[2] Department of Business Information Systems and Operations Management,
Belk College of Business, University of North Carolina at Charlotte,
9201 University City Blvd, Charlotte, NC 28223, USA
{kzhao2,rlkumar}@uncc.edu

Abstract. This study aims to examine the antecedents and consequences of IT Internal Control Weaknesses (ITICWs). Specifically, we propose a comprehensive model to examine the impact of IT governance on ITICWs as well as effects of ITICWs on firm performance. To gain deep insight into ITICWs, we propose to apply text mining analytics to categorize different types of ITICWs. This allows us to examine the impact of different categories of ITICWs on firm performance. We are in the process of collecting the data.

Keywords: IT governance · IT internal control weaknesses · Firm performance · Text mining

1 Introduction

With the fast pace of information technology (IT), modern organizations rely heavily on IT for their business operations. IT business value research suggests that IT creates firm value [5]. However IT generated business value often concur with risks such as data breach, which could result in negative impact on firm performance. These risks are not only technical problems of IT, but also internal control problems of management. Management should develop proper internal control procedures to prevent the generation of these risks and protect firms' assets [7].

Extant auditing standards (AS) have indicated the potential effect of IT on internal controls [8]. For instance, the statement on AS (SAS) No. 109 describes that the use of IT poses both benefits and risks to internal controls. In order to help reducing certain internal control risks, the Sarbanes-Oxley Act (SOX) was enacted by U.S. congress to set more rigorous standards to regulate the public compliance. Even though several authors summarize the literature on internal control weaknesses (ICWs) in U.S. [9] and study the relationships among IT governance, ITICWs, and financial performance [3], research on the antecedents and consequences of ITICWs is still limited.

© Springer International Publishing Switzerland 2016
V. Sugumaran et al. (Eds.): WEB 2015, LNBIP 258, pp. 155–160, 2016.
DOI: 10.1007/978-3-319-45408-5_13

Since ITICWs are one special category of ICWs, we are thus motivated to study the ITICWs in a comprehensive view. In this paper, we investigate the effective IT governance on ITICWs, and the impact of ITICWs on firm performance. Specifically, we will focus on which types of ITICWs impact firm performance.

2   Literature Review and Hypotheses Development

This study builds on two research streams within the literature: (1) the impact of IT governance on ITICWs [3]; (2) the association between ITICWs and firm performance [6]. We use the literature as the theoretical background to integrate the two streams and examine the antecedents and consequences of ITICWs. We first discuss the role of IT governance on ITICWs disclosure. Second, we review the impact of ITICWs on firm performance.

2.1   IT Governance and ITICWs

According to the prior literature, we expect IT governance to play an important role in ensuring the quality of a firm's IT internal controls. Effective IT governance over planning and the system development life cycle should result in more accurate and timely financial reporting [13]. In this study, we combine two research areas to construct an IT governance score based on secondary data. We identify and combine indicators from IT governance and leadership literature [1, 12]. We categorize our indicators into three groups (oversight, leadership IT background, and IT leadership importance) based on IT governance definition. We argue that firms with a stronger oversight function are more likely to supervise top managers in IT implementation and controls because "Part of audit committee's role is to look for ways to identify risk. In general, it becomes an independent guardian of the entity's assets." [11, p. 5]. Bassellier et al. (2003) stated that IT-related experience that executives possess enables them to exhibit IT leadership in their area of business. We believe that leaders with IT experience may respond to IT internal control weaknesses the company faces in a timely manner and remediate them appropriately.

H1:   IT governance is negatively associated with ITICWs

2.2   ITICWs and Firm Performance

Internal controls are built around firm objectives and are established to control any threats that might impact firm operations and performance. As IT internal controls are an important category of internal controls in today's computer-intensive business environment, it may also have an adverse effect on an organization. Firms without effective IT internal controls are more likely to have inefficient transaction process, inadequate systems, and insufficient access control, which might result in vulnerability of a firm's sensitive data that can lead to negative firm performance.

On December, 2006, TJX companies Inc. discovered an unauthorized intrusion into their computer systems, and determined that personal and confidential customer information was stolen. Several ITICWs were identified, such as the lack of logical and physical access control to the corporate network and operational system, which left private customer information vulnerable. The scope of the breach spanned about 18 months before it was detected, indicating a lack of timely monitoring or detection of internal control risks. The ITICWs lead to a huge financial impact with TJX spending $171.5 million pre-tax related to the computer intrusion, and maintains $42.2 million reserve for future losses related to the breach [7]. Therefore, we propose that firms with ITICWs are more likely to have worse performance.

H2: ITICWs is negatively associated with firm performance

3 Variable Definitions and Proposed Research Model

Based on the above hypotheses, firm performance is affected by firms' ITICWs disclosure, and IT governance plays a role in ITICWs discovery. The proposed research model is shown in Fig. 1.

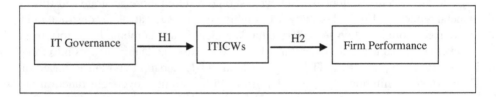

Fig. 1. The proposed research model

3.1 IT Governance Score

IT governance: Since IT governance is one aspect of corporate governance, we construct the IT governance score based upon the corporate governance literature [4]. We follow the Gov-Score proposed by them and construct the ITGOV-score. We compute the overall ITGOV-score as a summary of IT governance measure including 11 factors.

3.2 ITICWs Variable

Under the SOX 404, all publicly traded companies are mandated to disclose deficiencies in internal controls. The most severe type of internal control deficiencies is referred to as the ICWs. If the ICWs are IT-related, we refer to them as ITICWs.

Groups of ITICWs are detected using one of the text mining algorithms known as latent semantic analysis (LSA) or latent semantic indexing (LSI). LSA is well known for concept extraction, information retrieval, modeling human understanding of word meaning, and operating on textual data [10]. The detailed information of ITICWs detection is discussed in the section of data collection.

3.3 Firm Performance Variable

Return on assets (ROA) has been extensively used in prior IT value literature [2]. ROA identifies a firm's ability to generate productivity from its assets without regard to how they are financed, and it is a more comprehensive measure [3]. We thus use ROA to measure firm performance in this paper since ITICWs are linked to actual future earnings.

3.4 Variables' Measurement

Based on a review of prior studies, our model includes variables such as: ROA (Net income/total assets), ITICWs (Groups of ITICWs detected using text mining), ITGOV (Multiple items based on audit, composition of board, CxO IT experience, and CIO compensation), and control variables such as: SIZE (Firm size: the natural logarithm of the total assets of the firm.), AGE (Firm age: the log of the number of years the firm has CRSP data.), Earnings(Earnings before extraordinary income), SG (One-year sales growth rate: sales for firm j in year t/sales for firm j in year t-1), MB (Market-to-Book ratio: market valuation/book value of equity), ADV (Advertising expense/sales), R&D (Research and development expense/sales), and CAP (Capital expenditures/sales).

4 Research Method

4.1 Data Collection

To validate our model, we collect data from multiple sources including Compustat, CRSP, Mergent online, proxy statement, and 10-K filings. We start with the SOX 404 reports to identify the initial sample of the firms with ICWs. If the company has ICWs, we then determine whether the ICWs are IT-related. Once the ITICWs were identified and extracted in separated files, we further proceed to the use of LSA to categorize the ITICWs.

LSA starts with an initial term by document matrix that undergo a series of transformation followed by a singular value decomposition (SVD) transformation that generates term eigenvectors, square roots eigenvalues (known as singular values in descending order), and document eigenvectors. The term-by-dimension matrix shows the terms describing an extracted factor. The corresponding documents to each term are easily retrievable for the interpretation of the extracted factors. When multiply term eigenvectors by the singular values the result is a term-by-factor matrix of term loadings, and when multiply document eigenvectors by the singular values the result is a document-by-factor matrix of document loadings. By correlating factor terms with the corresponding document factors each factor is labeled.

Following the recommendations in the field [14], we extract the following factors: ineffective IT departments, inadequate audit trails, limited errors detected in spreadsheet, lack of segregation of duties, inadequate security and disaster recovery, inadequate accounting system procedures, lack of backup procedures, improper management personal access to the system, lack of methodological and monitoring activities, lack of general accounting system policies, lack of IT systems prevention and detection of fraud, ineffective IT security control, inadequate IT staffing, and lack of integrated complex accounting information systems.

5 Expected Contributions and Implications

Internal controls have become an important issue for both researchers and practitioners especially after the implementation of the SOX act. With the pervasive IT in modern organizations, IT internal controls have become increasingly critical for organizations to reduce risks, such as sensitive data loss, system and infrastructure threat, and transaction fraud. In this study, we propose a comprehensive model to examine the antecedent and consequence of ITICWs.

This study is expected to make several contributions to the growing literature. First, this study intends to examine whether IT governance helps to mitigate ITICWs. Second, this study examines the performance impact of ITICWs. Third, this study extends prior studies by using text mining to document ITICWs, and investigate the antecedents and consequences of ITICWs.

This study should also be of interest to professionals to guide firms in building stronger IT governance to implement effective IT internal controls, and result in better firm performance.

References

1. Bassellier, G., Benbasat, I., Reich, B.H.: The influence of business managers' IT competence on championing IT. Inf. Syst. Res. **14**, 317–336 (2003)
2. Bharadwaj, A.S.: A resource-based perspective on information technology capability and firm performance: an empirical investigation. MIS Q. **24**, 169–196 (2000)
3. Boritz, J.E., Lim, J.H.: IT control weaknesses, IT governance and firm performance, Working paper (2008)
4. Brown, L.D., Caylor, M.L.: Corporate governance and firm valuation. J. Account. Public Policy **25**, 409–434 (2006)
5. Brynjolfsson, E., Hitt, L.M.: Beyond computation: information technology, organizational transformation and business performance. J. Econ. Perspect. **14**, 23–48 (2000)
6. Carter, L., Phillips, B., Millington, P.: The impact of information technology internal controls on firm performance. J. Organ. End User Comput. **24**, 39–49 (2012)
7. Cereola, S.J., Cereola, R.J.: Breach of data at TJX: an instructional case used to study COSO and COBIT, with a focus on computer controls, data security, and privacy legislation. Issues Account. Educ. **26**, 521–545 (2011)
8. Chen, Y., Smith, A.L., Cao, J., Xia, W.: Information technology capability, internal control effectiveness, and audit fees and delays. J. Inf. Syst. **28**, 149–180 (2014)
9. El-Mahdy, D.F., Thiruvadi, S.: Antecedents, characteristics and consequences of internal control weaknesses and the COSO (2013) framework. Doctoral Dissertation (2014)
10. Evangelopoulos, N., Visinescu, L.L.: Text-mining the voice of the people. Commun. ACM **55**, 62–69 (2012)

11. Hall, A.J.: Information Technology Auditing, 3rd edn. Cengage Learning, South-Western (2011)
12. Li, C., Lim, J.-H., Wang, Q.: Internal and exteranl influences on IT control governance. Int. J. Account. Inf. Syst. **8**, 225–239 (2007)
13. Masli, A., Richardson, V.J., Sanchez, J.M., Smith, R.E.: Returns to IT excellence: evidence from financial performance around information technology excellence awards. Int. J. Account. Inf. Syst. **12**, 189–205 (2011)
14. Visinescu, L.L., Evangelopoulos, N.: Orthogonal rotations in latent semantic analysis: an empirical study. Decis. Support Syst. **62**, 131–143 (2014)

Creating Realistic Synthetic Incident Data

Nico Roedder[1]([✉]), Paul Karaenke[2], and Christof Weinhardt[3]

[1] Information Process Engineering,
FZI Research Center for Information Technology, Karlsruhe, Germany
roedder@fzi.de
[2] Department of Informatics,TU München, Garching, Germany
karaenke@in.tum.de
[3] Institute of Information Systems and Marketing,
Karlsruhe Institute of Technology, Karlsruhe, Germany
weinhardt@kit.edu

Abstract. The utilisation of the full flexibility of on-demand IT service provisioning requires in-depth knowledge on service performance. Otherwise reduction in cost going along with an increase of availability cannot be achieved. Thus, IT service decision methods incorporating IT service incident data are required. However, a lot of these models cannot be evaluated in a satisfactory fashion due to the lack of real-world incident data. To address this problem, we identify the need for realistic synthetic incident data for IT services. We stipulate the composition of this incident data and proclaim a procedure enabling the creation of realistic synthetic incident data for IT services allowing for a thorough evaluation of any formal decision model that relies on these forms of data sources.

Keywords: On-demand services · Cloud computing · Simulation · Evaluation · Decision support

1 Introduction

When services fail service management tends to be interested in answering three questions. First: How do we fix it fast? Second: Who is responsible? And Third: How do we stop it from happening again? It is a generally accepted fact that the most significant failures in information technology (IT) environments (as in other industry environments) are due to human errors [1–3]. Famous examples such as the June 29, 2001 NASDAQ integrity failure which was caused during the routine testing of a development system by an administrator, lead up to more recent service disruptions like the April 29, 2011 Amazon Web Services (AWS) 47 h downtime that started with an incorrectly performed network traffic shift by a network technician [4]. So what should be the consequence of this perception other than a twist on "What can go wrong will go wrong?" Failures are always an interaction of inchoate elements where even a small error can lead to a disaster. And it is hardly possible to foresee all risks even by the most formidable group of experts. With computational power ever increasing and analytical methods for

© Springer International Publishing Switzerland 2016
V. Sugumaran et al. (Eds.): WEB 2015, LNBIP 258, pp. 161–165, 2016.
DOI: 10.1007/978-3-319-45408-5_14

analysing data streams becoming better and better there is a soaring number of research that deals with the analysis of incident and monitoring data to improve stability on the one hand and support IT service decisions on the other hand.

The evaluation of these new models however, is very bothersome as there is hardly any usable real-world IT service incident or monitoring data available. A lot of research in cloud computing and decision support systems relies on a series of case studies or is tailored specifically for the scenario with a project partner who is (understandably) not willing to publish their service incident data. Other available data e.g. from cloud providers like Amazon Web Services[1] or Salesforce[2] is aggregated in a way that makes it unusable for most cases. Consequently there is a need for the creation of synthetic incident data.

We thus, state the following: it should be the aim of any researcher requiring service monitoring or incident data to validate his work with real-world data. Unfortunately this data is very hard to come by. Or to specify: almost impossible to come by with certain necessary properties to make it comparable.

The goal of this research is the identification of characteristics of a procedure to create realistic, comparable and reproducible incident data to validate formal models from the realm of service science research.

This work is structured as follows: In Sect. 2 we analyse related works. Section 3 summarises the requirements for realistic incident data and proclaims its characteristics. We conclude our work in Sect. 4.

2   Literature Review

As stated in the section before, it should be the aim of a researcher to validate his work with real-world data if there is any way to do so. Because of this established fact, research dealing with the creation of artificial incident data is sparse. Nevertheless there has been significant research on identifying the failure patterns of service incidents. Franke has analysed empirical data sets and concluded that the Weibull distribution and the Log-normal distribution are suited for a fitting in [5,6].

When exploring a connection between business impact costs and service incidents Kieninger et al. have identified the Beta distribution to fit their empirical data sets [7,8]. The focus is on finding the relation between these incidents and business costs.

Google researchers have conducted analyses on the failure of hard drives in their data centres [9]. However, they stick with exhibiting the results of their findings without trying to fit them to probability distributions. When scoping these results it can be assumed that disk failures are somewhat normally distributed.

While these findings are very interesting indeed, the aim is never to reproduce these incident patterns for future experiments and simulation studies. They

---

[1] http://status.aws.amazon.com/.
[2] https://trust.salesforce.com/trust/status/.

should serve as a basis when deciding for the correct distributions or rather the incident patterns that should be analysed.

3   Model Description

We first identify reasons why real-world monitoring data is not sufficient or to hard to come by when dealing with decisions in IT service settings. The subsection thereafter comprises characteristics necessary for simulated services.

3.1   The Need for Synthetic Incident Data

In this section we want to identify why there is a need for incident data to be synthetic when validating decision models relying on IT service incident and monitoring data.

Disposal and Aggregation. Most service providers dispose of their monitoring and incident data as soon as it is no longer needed for contractual obligations or internal analysis. In most cases fine granulations of monitoring data is deleted after a set time period and only aggregated data is stored over a longer course of time. However, smaller service providers might not even keep this aggregated data.

Service Changes. In realistic settings IT infrastructures change over the course of time. A provider might increase computing power, change other hardware components or improve/change the software running on the infrastructure. This makes fair comparisons impossible.

Different Parameters. Besides their functionality, IT services have certain parameters that are non-functional (e.g. availability). Comparisons of these parameters be only conduced when extracting them to a common format (e.g. WS-Agreement [10]) and reducing the set of parameters to the ones available for all services.

Segment Lengths. Depending on what kind of service is monitored, the time segment length might significantly differ. Some services are monitored on a millisecond basis, while the availability of other services is tested once an hour. This also makes comparisons between services assailable.

These limitations are most prominent when dealing with real-world incident data and create the need for synthetic data. For this data to be realistic and be of use in IT service decision scenarios a series of characteristics have to be simulated. These are listed in the following section.

3.2   Service Characteristics

It is assumed that IT services have unique generic types i (e.g. storage or database service) and are offered by multiple service providers j. Specific services are the combination of an IT service type and a provider offering that service s_{ij}. Each service has a price p_{ij} per unit of time t it is contracted.

IT service incidents have a frequency $m_f(s_{ij}, t)$ and an expected failure duration $d_f(s_{ij}, t)$. This is common practice in reliability engineering where the frequency is often labelled as the *Rate Of Occurrence Of Failures* (ROCOF) and the failure duration as *Mean Time To Repair* (MTTR) [11]. Both are substituted to $\lambda_{ij}^t := (m_f(s_{ij}, t), d_f(s_{ij}, t)) \in \Lambda$. Some services s_{ij} come with a penalty agreement $\mu_{ij}(\cdot)$ in case service objectives are not met. The penalty that has to be paid by the service provider is dependent on λ_{ij}^t.

For incident data to be realistic in a service decision scenario the afore introduced components have to be simulated.

Pricing Strategy. IT services tend to be priced in tiers. Providers offer e.g. gold and platinum plans, where the platinum plan is significantly more expensive than the gold plan and offers a higher quality. Additionally usage-based pricing, performance-based pricing, user-based pricing and flat pricing should be implemented [12,13].

Penalty Agreements. Especially public cloud computing providers offer no penalty payments when a service is unavailable. Traditional (outsourcing) service providers however are contractually obligated to pay a fee if their service is not usable. In reality, however, only a series of functions are encountered that are generally limited by an upper bound [14,15].

Service Failures. It is assumed that services fail with certain probabilities that can be approximated through failure distributions. This is a well-established fact for systems in reliability engineering research [16,17], and also valid across the wide range of different IT services, no matter if considering human error [2], hardware failures [9] or other unanticipated failures [6,8].

Each of the above characteristics are necessary for a simulative comparison of different IT services and their incident behaviour. The correct parametrisation of different failure distributions is vital for meaningful incident time series.

4   Conclusion

In this work we have primarily shown that there is a need for synthetic incident data. Having evaluated a previously introduced decision method [18–20], we want to improve an existing implementation significantly and provide a thorough survey of the generated data and package our model and make it available for usage. Hence, the work at hand is conducted as research in progress to gather further input from fellow researchers and enhance our model.

References

1. Kirwan, B.: Human reliability assessment. Encyclopedia of Quantitative Risk Analysis and Assessment (2008)
2. Reason, J.: Human error: models and management. Bmj **320**(7237), 768–770 (2000)
3. Ayachitula, N., Buco, M., Diao, Y., Maheswaran, S., Pavuluri, R., Shwartz, L., Ward, C.: IT service management automation-a hybrid methodology to integrate and orchestrate collaborative human centric and automation centric workflows. In: IEEE International Conference on Services Computing, SCC 2007, pp. 574–581. IEEE (2007)

4. AWS-Team: Summary of the Amazon EC2 and Amazon RDS Service Disruption in the US East Region (2011). http://aws.amazon.com/message/65648/. Accessed 09 Nov 2015
5. Franke, U., Holm, H., König, J.: The distribution of time to recovery of enterprise IT services. IEEE Trans. Reliab. **63**(4), 858–867 (2014)
6. Franke, U.: Optimal IT service availability: shorter outages, or fewer? IEEE Trans. Netw. Serv. Manag. **9**(1), 22–33 (2012)
7. Kieninger, A., Straeten, D., Kimbrough, S.O., Schmitz, B., Satzger, G.: Leveraging service incident analytics to determine cost-optimal service offers. In: Wirtschaftsinformatik, 64 (2013)
8. Kieninger, A., Berghoff, F., Fromm, H., Satzger, G.: Simulation-based quantification of business impacts caused by service incidents. In: Falcão e Cunha, J., Snene, M., Nóvoa, H. (eds.) IESS 2013. LNBIP, vol. 143, pp. 170–185. Springer, Heidelberg (2013)
9. Pinheiro, E., Weber, W.-D., Barroso, L.A.: Failure trends in a large disk drive population. In: FAST, vol. 7, pp. 17–23 (2007)
10. Andrieux, A., Czajkowski, K., Dan, A., Keahey, K., Ludwig, H., Nakata, T., Pruyne, J., Rofrano, J., Tuecke, S., Xu, M.: Web Services Agreement Specification (WS-Agreement). Open Grid Forum (OGF) Proposed Recommendation GFD.107 (2007)
11. Yeh, L.: The rate of occurrence of failures. J. Appl. Probab. **34**(1), 234–247 (1997)
12. Harmon, R., Demirkan, H., Hefley, B., Auseklis, N.: Pricing strategies for information technology services: a value-based approach. In: 42nd Hawaii International Conference on System Sciences, HICSS 2009, pp. 1–10. IEEE (2009)
13. Kesidis, G., Das, A., de Veciana, G.: On flat-rate and usage-based pricing for tiered commodity internet services. In: 42nd Annual Conference on Information Sciences and Systems, CISS 2008, pp. 304–308, March 2008
14. Moon, H.J., Chi, Y., Hacigumus, H.: SLA-aware profit optimization in cloud services via resource scheduling. In: 2010 6th World Congress on Services (SERVICES-1), pp. 152–153. IEEE (2010)
15. Buco, M.J., Chang, R.N., Luan, L.Z., Ward, C., Wolf, J.L., Yu, P.S.: Utility computing SLA management based upon business objectives. IBM Syst. J. **43**(1), 159–178 (2004)
16. Elsayed, E.A.: Reliability Engineering, vol. 88. Wiley, Hoboken (2012)
17. Kapur, K.C., Pecht, M.: Reliability Engineering. Wiley, Hoboken (2014)
18. Roedder, N., Knapper, R., Martin, J.: Risk in modern IT service landscapes: towards a dynamic model. In: 2012 5th IEEE International Conference on Service-Oriented Computing and Applications (SOCA), pp. 1–4, December 2012
19. Roedder, N., Karaenke, P., Knapper, R.: A Risk-aware decision model for service sourcing (Short Paper). In: 2013 IEEE 6th International Conference on Service-Oriented Computing and Applications (SOCA), pp. 135–139, December 2013
20. Roedder, N. and Karaenke, P. and Knapper, R. and Weinhardt, C.: Decision-making based on incident data analysis. In: 2014 IEEE 16th Conference on Business Informatics (CBI), vol. 1, pp. 46–53, July 2014

Rumor and Truth Spreading Patterns on Social Network Sites During Social Crisis: Big Data Analytics Approach

Mehrdad Koohikamali[1,2(✉)] and Dan J. Kim[2]

[1] School of Business, University of Redlands, Redlands, CA, USA
mehrdad_koohikamali@redlands.edu
[2] ITDS Department, College of Business, University of North Texas, Denton, TX, USA
Dan.Kim@unt.edu

Abstract. Social network sites give their users the ability to create contents and share it with others. During social crisis, the spread of false and true information could have profound impacts on users. Lack of prior studies to compare differences between diffusion patterns of rumors and truths during social crisis is the motivation of this study. In this study, we examine the role of information credibility, anxiety, personal involvement, and social ties on rumor and truth spread during social crisis. Building on the rumor theory, we propose a research model to examine differences between spread of rumors and truths. Using the Tweeter data collected during the Baltimore riots in 2015, we test the research model. Theoretical contributions and practical implications will be outlined based on the findings of the study. We anticipate findings will provide new avenues of research by determining characteristics of truths and rumors in online contexts.

Keywords: Rumor and truth in social crisis · Spread patterns of rumor and truth · Big data analytics · Information diffusion

1 Introduction

Social network sites (SNSs) provide a rich medium to generate content and share it with other users. People use SNSs for various reasons such as communicating, information seeking, maintaining close ties, building identity, information sharing, location disclosing, and educating [1, 2]. Understanding rumors diffusion pattern in online environments has been the focus of many studies in recent years [3]. However, isolating rumors from truths can be misleading without considering the spread of true information [4]. The way rumors spread is influenced by the absence of truths and truth presence could affect rumor diffusion.

During social crisis and disasters, the uncertainty of information related to the event accompanied by the public anxiety and it increases negative consequences of rumors spread. Recent social crisis and events such as the Baltimore riots in 2015 indicate sharing information on social media has profound influence on involved subjects. In previous research, there is a gap to explore the differences between diffusion patterns of rumors and truth in social crisis and disaster situations on SNSs. This research gap,

© Springer International Publishing Switzerland 2016
V. Sugumaran et al. (Eds.): WEB 2015, LNBIP 258, pp. 166–170, 2016.
DOI: 10.1007/978-3-319-45408-5_15

specifically in the areas of rumor and truth diffusion on SNSs, leads this research to pose several research questions. In the SNS environment, we propose two questions:

- What are the truth and rumor diffusion patterns during social crisis on SNSs?
- How does the truth diffusion pattern is different from rumor?

2 Literature Review

2.1 Information Diffusion on SNSs

People use SNSs for different purposes. According to prior research many people are aware of their intentions to use SNS. SNSs provide tools to create and share contents. SNSs have changed the way information is generated, distributed, and shared among many societies [5]. A micro-blogging service such as Twitter is mainly used for rapid information dissemination during social crises [5]. Rumors can spread in many contexts, but they are more prevalent in uncertain situations [3]. During the first stage of a social crisis many unproven facts should be rejected (rumors) or accepted (truths). During disasters people tend to fill in blanks, improvise news, and spread rumors [3].

2.2 Rumor and Truth Spread on SNSs

Rumor has different meaning in different contexts. Unverified propositions for beliefs related to the topic of interest and uncertain truths about an involved subject are known as rumors [4]. Some researchers believe rumors are claims of facts about people, groups, events, and institutions without any proof of being true [6]. Rumor spreads in three phases: rumor parturition, rumor diffusion, and rumor controlling [3]. Rumor spread pattern occurs in two general forms. First, in social chain pattern a rumor moves from one person to another person with single interaction [7]. The second form of rumor spread pattern, known as multiple interaction network, and users receive a rumor from different users and send it to others [7].

2.3 Information Credibility

Information quality is a multi-dimensional construct, and several models for the conceptualization and measurement of it exist. Research suggests that information quality is highly situational and depends on the source and content of information [8]. We adopt the information quality framework in the context of online reviews. In the context of online customer reviews, identifying true and false information is possible if readers have actual experience with products/services [8]. While during social crisis the rumor and truth are not easily separable due to the lack of enough evidences. In this study, for the content quality and reputation for source quality we adopt following dimensions: believability, relevancy, completeness, and concise representation [9].

3   Theoretical Background

Rumor theory explains the influence of ambiguous situations on spread of rumors [10]. During social crisis, many individuals' beliefs are influenced by rumors because people tend to seek others' opinions in such circumstances [11]. The theoretical foundation of this research is based on the rumor (mongering) model [5]. The rumor (mongering) model suggests rumors spread due to ambiguity of information, anxiety, personal involvement, and social ties [5].

Information credibility represents the quality of information. During crisis, the lack of information creates ambiguous and uncertain conditions [12]. Information credibility might reflect the content quality or the source quality. Content quality can be explained by believability, relevancy, completeness, and concise representation [9]. If the content of information is credible there is less doubt to correctly interpret the message [5]. Online information credibility is often judged based on the content [13]. In addition, credible sources are often believed to spread true information. During social crisis uncertain conditions provide both the source and content to influence the spread of rumors and truths. We propose,

H1a: Information credibility positively influences the spread of true information.

H1b: Information credibility positively influences the spread of rumors.

Anxiety is a multidimensional construct and it can be explained as trait anxiety or state anxiety phenomena [14]. Uncertain conditions of social crisis cause higher anxiety among users [15]. Anxious people are more likely to seek information [16]. Moreover, In a SNS environment people are able to seek truthful information and subsequently lessen their anxiety. Likewise, we believe SNS users spread truths during uncertain conditions to soothe others as well as themselves. Therefore, we hypothesize:

H2a: Anxiety positively influences truth spread.

H2b: Anxiety positively influences rumor spread.

The perceived importance of a social crises for a user plays an important role in rumor spread phenomena [3]. Expressions of personal involvement in rumors is positively related with rumor spread [3]. During social crisis on SNSs, people transmit more rumors if they feel they are involved [5]. Even though people want to refrain to pass rumors, they pass more rumors during crisis. Contrary to rumor spread, the truth spread is negatively influenced by personal involvement. We propose:

H3a: Feelings of personal involvement negatively influences truth spread.

H3b: Feelings of personal involvement positively influences rumor spread.

Previous research show the relationship between existence of social ties and rumor spread [5]. Social ties are based on an existing social relation and the message is received from a known person has a greater chance of spreading [5]. Directed messages reflect strong ties if there is a two-way relationship and weak social ties do not influence the rumor spread. Similarly, the strength of social ties could be related with the spread of truth information due to the trustiness between ties. Thus, we hypothesize:

H4a: Strong social ties positively influence truth spread on SNSs.

H4b: Strong social ties positively influence rumor spread on SNSs.

Following the rumor theory and proposed hypotheses, the posited research model is illustrated in Fig. 1.

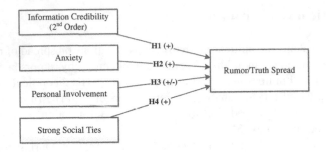

Fig. 1. The proposed research model.

4   Methodology

To test and verify the proposed model a combination of coding procedure and machine learning algorithm of text mining will be used. The machine learning algorithm is based on the Koohikamali and Kim [17] study. We collected a Twitter dataset of Baltimore riots in 2015. During the period of the crisis from April 18, 2015 to May 03, 2015 we collected more than three hundred thousand tweets related to the Baltimore riots event. It includes a good example of social crisis that people used SNSs such as Twitter to generate and seek information ranging from calls to protest to pleas for prayer.

5   Expected Results and Contributions

This study will potentially contribute to both theory and practices in the area of social networking and information sharing. To best of our knowledge, this is the first study to examine the differences between the truth and rumor spread during rumor diffusion phase. We anticipate our study to contribute to the body of knowledge in many research areas by providing important aspects of rumor diffusion when truth is present. Results can be influential for research in communication, marketing, media, e-commerce, and information systems. The second contribution of this study is its emphasis on how information credibility influences the rumor and truth spread. We believe results of this study will broaden the current knowledge about online information dissemination.

6   Conclusion

This study serves as initiative attempt to investigate the differences between truth and rumor spreads on SNSs during social crisis. In this study we provided a research model to investigate the differences between rumor and truth spread patterns. It is built on the theory of rumor to explain effects of information credibility, anxiety, personal involvement, and strong social ties on rumor and truth spread.

References

1. Kane, G.C., et al.: What's different about social media networks? A framework and research agenda. MIS Q. **38**(1), 275–304 (2014)
2. Koohikamali, M., Gerhart, N., Mousavizadeh, M.: Location disclosure on LB-SNAs: The role of incentives on sharing behavior. Decis. Support Syst. **71**, 78–87 (2015)
3. Liu, F., Burton-Jones, A., Xu, D.: Rumors on social media in disasters: extending transmission to retransmission. In: Pacific Asia Conference on Information Systems (2014)
4. Berinsky, A.J.: Rumors, Truths, and Reality: A Study of Political Misinformation. Massachusetts Institute of Technology, Cambridge (2012). Manuscript
5. Oh, O., Agrawal, M., Rao, H.R.: Community intelligence and social media services: a rumor theoretic analysis of tweets during social crises. MIS Q. **37**(2), 407–426 (2013)
6. Einwiller, S.A., Kamins, M.A.: Rumor has it: the moderating effect of identification on rumor impact and the effectiveness of rumor refutation1. J. Appl. Soc. Psychol. **38**(9), 2248–2272 (2008)
7. Buckner, H.T.: A theory of rumor transmission. Publ. Opin. Q. **29**(1), 54–70 (1965)
8. Mousavizadeh, M., Koohikamali, M., Salehan, M.: Antecedents of online customers reviews' helpfulness: a support vector machine approach. In: International Conference on Information Systems. Fort Worth (2015)
9. Chen, C.C., Tseng, Y.-D.: Quality evaluation of product reviews using an information quality framework. Decis. Support Syst. **50**(4), 755–768 (2011)
10. Oh, O., Kwon, K.H., Rao, H.R.: An exploration of social media in extreme events: rumor theory and twitter during the Haiti earthquake 2010. In: ICIS (2010)
11. Chen, H., Lu, Y., Suen, W.: The power of whispers: a theory of rumor, communication and revolution. In: 2013 Meeting Papers. Society for Economic Dynamics (2013)
12. Smart, P.R., Sycara, K.: Cognitive social simulation and collective sensemaking: an approach using the ACT-R cognitive architecture. In: The Sixth International Conference on Advanced Cognitive Technologies and Applications (2014)
13. Flanagin, A.J., Metzger, M.J.: Perceptions of internet information credibility. J. Mass Commun. Q. **77**(3), 515–540 (2000)
14. Sharp, P.B., Miller, G.A., Heller, W.: Transdiagnostic dimensions of anxiety: neural mechanisms, executive functions, and new directions. Int. J. Psychophysiol. **98**(2), 365–377 (2015)
15. Lee, J., Agrawal, M., Rao, H.: Message diffusion through social network service: the case of rumor and non-rumor related tweets during Boston bombing. Inf. Syst. Front. **2015**, 1–9 (2013)
16. Gino, F., Brooks, A.W., Schweitzer, M.E.: Anxiety, advice, and the ability to discern: feeling anxious motivates individuals to seek and use advice. J. Pers. Soc. Psychol. **102**(3), 497–512 (2012)
17. Koohikamali, M., Kim, D.: Does information sensitivity make a difference? Mobile applications' privacy statements: a text mining approach. In: Americas Conference on Information Systems. AIS, Puerto Rico (2015)

The Big Data Analysis of the Next Generation Video Surveillance System for Public Security

Zheng Xu[1,2(✉)], Zhiguo Yan[1], Lin Mei[1], and Hui Zhang[2]

[1] The Third Research Institute of the Ministry of Public Security, Shanghai, China
xuzheng@shu.edu.cn
[2] Tsinghua University, Beijing, China

Abstract. Video surveillance has become the main tool due to its rich, intuitive and accurate information. However, with the large-scale construction of video surveillance systems all over the world, problems such as "useful information and clues cannot be found immediately with video big data" decrease detecting efficiency during crime prediction and public security governance. This paper concludes a novel architecture for next generation public security system, and the "front+back" pattern is adopted. Under the architecture, cloud computing technologies such as distributed storage and computing, data retrieval of huge and heterogeneous data are introduced, and multiple optimized strategies to enhance the utilization of resources and efficiency of tasks.

Keywords: Big data · Public security · Data retrieval · Video surveillance system

1 Introduction

Recently, the worldwide terrorist incidents and crime events occur frequently, and it is urgent for governments and police to pay serious attention to the public security governance, the traffic accidents, criminal prediction and other incidents. With the help of cloud computing [1, 2], internet of things [3, 4], and Big Data [5, 6], video surveillance has become the main tool due to its rich, intuitive and accurate information. A great amount of video surveillance systems have been built all over the world. China has built more than 23 million video surveillance cameras till 2013, of which 3 million are utilized by police, and the video surveillance are entering the big data era with its 4 V properties. That is, the video data has very huge volume, taking one city for example, thousands of cameras are built of which each collects high-definition video over 24 to 48 GB every day with the rapidly growth; secondly, data collected includes variety of formats involving multimedia, images and other unstructured data; furthermore the valuable information contains in only a few frames called key frames of massive video data; and the last problem caused is how to improve the processing velocity of a large amount of original video with computers, so as to enhance the crime prediction and detection effectiveness of police and users. Moreover, a great variety of public security information systems have been built, which have played important roles in the traffic accidents governance, crimes events and terrorist incidents prediction. Series of problems appear, on the one hand, redundant

V. Sugumaran et al. (Eds.): WEB 2015, LNBIP 258, pp. 171–175, 2016.
DOI: 10.1007/978-3-319-45408-5_16

construction of systems leads to great waste of resource, such as the video surveillance systems throughout the country, which are built with their independent software and hardware in each place.

To solve those problems, technologies such as knowledge mining and deduction, pattern recognition and cloud computing are widely utilized in the next generation video surveillance system, to assist police to discover valuable information and predict crime from large amount data. China National Laboratory of Pattern Recognition (NLPR) has developed the distributed video surveillance system [7], which is applied to discover unusual behavior and traffic violations with pattern recognition. The Industrial Technology Research Institute of Taiwan set up the Surveillance Video Analysis Center and built the "Cloud Intelligent Video Analysis and Retrieval System", which provides video retrieval and other video analysis services to aid police to discover crime efficiently.

2 Problems Description

Surveillance video data has the 4 V properties of big volume, variety of data format, low value and slow processing velocity, resulting in several problems in application especially in crime detection and public security management for police:

(1) Little video analysis technology is utilized to recognize vehicle information (license-plate, logo, and colour etc.) and simple applications, however crime prediction and clue discovery from massive video data are most rely on human detection, and it is still hard to discover deep information and complex content by computers, also lack of the standardized description of analysed content.
(2) Due to lack of effective resource management and organization, a great amount of computing and storage resources could not be utilized effectively when analyse and process video big data.
(3) Without the policing repository database, it is hard to mine the more complex relationship and deeper semantics from a great amount of data, also unnecessary to recommend police available information, clues, case trend.

3 Big Data Based Framework

We propose a new framework to show how to process, organize, manage and store massive video data. As shown in Fig. 1 The framework has three parts: video intelligent analysis (Object detection, target tracking, behaviour analysis and event analysis) and video structured description (VSD) are utilized to mine valuable information (persons, cars, unusual behaviours etc.) from large scale video data, which then is expressed in standard format. The second part is construction of policing repository database, which is used to data mining, information describing, moreover knowledge reasoning as the domain knowledge, and provide real cases to assist crime prediction. Furthermore virtualization and cloud computing provide efficient computing environment for techniques all above, and storage environment for various types of structured and unstructured data.

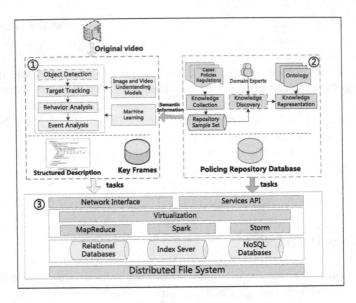

Fig. 1. The framework for the next generation video surveillance system

Video intelligent analysis and structured description are applied to deal with original video, of which the results are frames containing persons and cars, and their structured description with standard format. All the data are packaged with unified standard format and transferred to the distributed cloud platform which provides greatly efficient storing and computing ability. Due to the limited bandwidth, the "front+back" pattern is adopted, that is: simple video analysis algorithms are carried out in the cameras, and results are sent back to "the cloud" to support more complex computing and applications. The pattern could avoid network congestion caused by large-scale video big data.

Repository database could be constructed as follow steps: first, knowledge collection, that is collecting and analyzing existed cases, policies and regulations, and make them as knowledge repository sample set; secondly is knowledge discovery, that domain knowledge are mined, clustered and analyzed from the collected cases and rules, with machine learning such as support vector machines (SVM), or expert guidance; thirdly, knowledge representation, domain knowledge and rules should be represented with unified form such as RDFS, OWL and SWRL, and stored in repository database and model database, from which the information would be utilized to support training models, semantic retrieval, reasoning and crime prediction.

4 Big Data Analysis

We proposed the architecture of cloud platform for the next generation public security system, to show how to process, organize, manage and store large-scale heterogeneous data.

Due to the limited bandwidth, the "front+back" pattern is adopted, that is: data such as video, audio and other unstructured and structured data are collected by sensors such as cameras or from existed information systems and preprocessed in the "front" part, then the results are packaged with unified standard format and transferred to the "back" data center with strong storing and computing ability to support more complex computing and applications. The pattern could avoid network congestion caused by distributed heterogeneous data after data preprocessing in the "front"; in addition, based on the cloud computing and virtualization technologies, the cloud center realizes the resource consolidation of multiple IT resources, and provides unified computing and storage environment for more data analysis and applications such as data mining and semantic reasoning.

Take the video surveillance systems for example, video data are collected by cameras, in which ARM-based processing devices are embedded to do preprocessing such as video encoding, license-plate recognition, colour recognition under semantic description models, with the results including pictures and structured description with standard format transferred to the back center, and the original video data stored into databases deployed near cameras. The center provides more resources to support deep data analysis and applications for police. Data collected from types of public security systems are huge and heterogeneous, which brings great challenges for efficient storage and organization, fast retrieval and computing of data. Cloud computing technologies such as virtualization, distributed storage and computing are applied to solve these problems.

5   Conclusion

In this paper, we conclude a novel architecture for next generation public security system, and the "front+back" pattern is adopted to address the problems brought by the redundant construction of current public security information systems which realizes the resource consolidation of multiple IT resources, and provides unified computing and storage environment for more complex data analysis and applications such as data mining and semantic reasoning. Under the architecture, we introduce cloud computing technologies such as distributed storage and computing, data retrieval of huge and heterogeneous data, provide multiple optimized strategies to enhance the utilization of resources and efficiency of tasks. However, some other problems still exist: in what way the services could be provided to users, and it still cannot satisfy the routine detection and application for police. For example, combining crime prediction results with visualization methods is necessary for users during detection. These unsolved problems particularly merit our further study.

Acknowledgements. This work was supported in part by the National Science and Technology Major Project under Grant 2013ZX01033002-003, in part by the National High Technology Research and Development Program of China (863 Program) under Grant 2013AA014601, in part by the National Science Foundation of China under Grant 61300202, 61300028, in part by the Project of the Ministry of Public Security under Grant 2014JSYJB009, in part by the China

Postdoctoral Science Foundation under Grant 2014M560085, and in part by the Science Foundation of Shanghai under Grant 13ZR1452900.

References

1. Liu, Y., Zhu, Y., Ni, L., Xue, G.: A reliability-oriented transmission service in wireless sensor networks. IEEE Trans. Parallel Distrib. Syst. **22**(12), 2100–2107 (2011)
2. Liu, Y., Zhang, Q., Ni, L.: Opportunity-based topology control in wireless sensor networks. IEEE Trans. Parallel Distrib. Syst. **21**(3), 405–416 (2010)
3. Hu, C., Xu, Z., et al.: Semantic link network based model for organizing multimedia big data. IEEE Trans. Emerg. Topics Comput. **2**, 376–387 (2014)
4. Luo, X., Xu, Z., Yu, J., Chen, X.: Building association link network for semantic link on web resources. IEEE Trans. Autom. Sci. Eng. **8**(3), 482–494 (2011)
5. Xu, Z., et al.: Knowle: a semantic link network based system for organizing large scale online news events. Future Gener. Comput. Syst. **43**, 40–50 (2015)
6. Xu, Z., Luo, X., Zhang, S., Wei, X., Mei, L., Hu, C.: Mining temporal explicit and implicit semantic relations between entities using web search engines. Future Gener. Comput. Syst. **37**, 468–477 (2014)
7. Zhang, H., Mei, L., Liang, C., Sha, M., Zhu, L., Wu, J., Wu, Y.: Video structured description: a novel solution for visual surveillance. In: Qiu, G., Lam, K.M., Kiya, H., Xue, X.-Y., Kuo, C.-C., Lew, M.S. (eds.) PCM 2010, Part II. LNCS, vol. 6298, pp. 629–636. Springer, Heidelberg (2010)

Wish Lists and Shopping Carts: A Study of Visual Metaphors on Arab E-Commerce Websites

Divakaran Liginlal[✉], Maryam Al-Fehani, Preetha Gopinath, and Alex Cheek

Carnegie Mellon University Qatar, Education City, PO Box 24866 Doha, Qatar
{liginlal,malfehani,preethag,alexcheek}@cmu.edu

Abstract. A review of 3065 Arab e-commerce websites revealed that Web designers mostly adopt Western-influenced interface metaphors and pay scant attention to creating Arab culture-specific visual metaphors. In this research, we first collected and analyzed Arabic stories to extract themes for the creation of culturally attuned visual metaphors for use as Web interface elements. After considering various plausible themes, we specifically focused on the "wish list" metaphor as the object of our study. A subsequent eye-tracking experiment based on alternative designs of a shopping website provided evidence that participants paid more attention to the localized elements of website design. A questionnaire-based study further confirmed that localization of site design in general, and the use of culturally attuned visual metaphors in particular, gains the attention of Arab e-shoppers. The results provide both practitioners and researchers deeper insights into how culture shapes the design of websites.

Keywords: e-commerce · Culture · Visual metaphors · Icons · Eye-tracking

1 Introduction

Many researchers in the information systems discipline have studied the influence of culture on website design [1–3]. One common approach to measuring the cultural influence in the design of a website is to look at cultural markers. Khanum et al. [4] define cultural markers as "interface design elements that are largely influenced by cultural values." Our literature review confirms that the use of metaphors as cultural markers that influence the design of e-commerce websites has attracted little research interest. As Lakoff and Johnson [5] succinctly put it, "the essence of metaphors is understanding and experiencing one kind of thing in terms of another." A visual metaphor can be defined as the use of a visual element—image—(from one domain) representing another visual element from a different domain [6].

A study by Shen et al. [7] examined culture-centered design specifically in terms of Chinese culture. They applied a Chinese garden metaphor to create a Web browsing experience. Another study that explored culturally attuned user interface icons was by Heukelman and Obono [8]. Their study examined the possibility of an alternative to office metaphors by using an African village metaphor. These studies are examples of research targeted at Chinese and African cultures. Our literature search confirms that very few studies have looked at Arab e-commerce websites and culturally attuned visual metaphors [9].

© Springer International Publishing Switzerland 2016
V. Sugumaran et al. (Eds.): WEB 2015, LNBIP 258, pp. 176–180, 2016.
DOI: 10.1007/978-3-319-45408-5_17

2   Research Model and Methodology

Our research aims to answer one primary question: Does the use of visual metaphors that relate to Arab culture enhance the experience of Arab shoppers on e-commerce websites? In simple terms, the research aims to discover if the integration of Arab culture and visual metaphors influences Arab users of e-commerce websites.

We gathered thirty-three stories from books and from older citizens knowledgeable of Arab folklore and analyzed them to identify live characters, imaginary creatures, and inanimate objects. About 100 corresponding images were then collected from many sources on the Web to distill features that facilitated the creation of visual metaphors. Figure 1 shows examples of culturally adapted shopping cart icons and culturally adapted designs of the wish list icon conforming to the theme of Aladdin's lamp.

Fig. 1. Examples of icons resulting from the creative design process

Eye-tracking involves the process of measuring where an individual's eyes are focused at any given time, i.e., the gaze point, and the sequence of movement of the eyes from one point to another (saccades) [10]. Because eye-tracking alone may not help explain why a user is looking at a specific object or point in an image, our approach combines both eye-tracking and questionnaire studies.

Two alternative designs shown in Fig. 2 were considered for an eye-tracking experiment. For the Western-influenced design, the logo is simply the name of the business, Souq, which means shop. The icons and image are commonly used Western-influenced banners and product images and interface icons. For the Arab-influenced design, the logo is the word Souq written in Arabic, and the women and men icons represent Arab culture because they are wearing traditional Arab clothes. The wish list icon is portrayed with a magical lamp drawing inspiration from the popular Aladdin folklore.

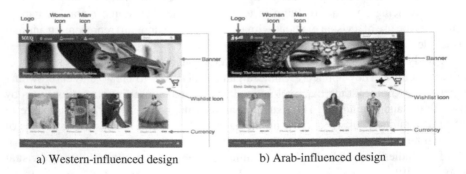

a) Western-influenced design   b) Arab-influenced design

Fig. 2. Home page design of e-commerce website

The interview questions were designed to identify features of the website that attracted the participants the most, the images or icons they could recall, and their specific and general impressions of the visual imagery of the site. Responses to another question that attempted to measure the perceived cultural affinity of the participant toward the viewed website were coded in such a way as to elicit a binary yes/no response. The participants were also required to retrospectively list the things on the website (pictures, icons, text) that made the website culturally friendly to them.

The study was conducted at Carnegie Mellon University's Qatar campus. All target participants were university students of Arab nationalities, age 18 or older, and native Arabic speakers. Because of the high resource requirements for eye-tracking studies, we reduced our number of participants to 10 for each design. The male-female ratio was 1:3. After viewing a warm-up website to help the participant adjust to viewing a website with the eye-tracker, the participant was asked to browse and navigate through the website to complete a shopping task. After completing the task, each participant answered the questionnaire containing 10 short questions about the website. Each study was completed in less than an hour.

3 Results and Discussion

Tobii Studio generates heat maps that represent the different areas of a screen where a user has looked longest. Orange and red areas represent the most time focused on an area of the screen, and green and yellow show those screen areas watched the least. The heat map shown in Figs. 3(a) and (b) clearly indicate that the wish list icon of the Arab-influenced design received the most attention relative to that on the Western design. Table 1 lists the mean values of the metrics generated from the eye-tracking data.

Fig. 3. Heat maps on home page of (a) Western-influenced site (b) Arab-influenced site

The time from the start of the stimulus display until the test participant fixates on the wish list icon for the first time was measured by the time-to-first-fixation metric. The mean values are 7.858 s and 5.698 s, respectively for the Western and Arab-influenced designs. Table 1 also shows the results of the student t-test for related comparison of means. The p-value of 0.18 and the t-value of 0.93717 are not high enough to assert statistically significant difference in the means. A fixation occurs when a participant's eye movement pauses on a specific area of the webpage, in this case the wish list icon.

The total fixation duration is the duration of all fixations on this area of interest. The mean total fixation duration of 1.155 s for the wish list icon in the Arab design is significantly greater than that of the mean total fixation duration of 0.47 s on the alternative design (t-value of −2.2969; p-value = 0.0169). Therefore, it appears that participants seemed to focus longer on the culturally attuned wish list icon, indicating greater interest.

Table 1. Key metrics from eye-tracking of the wish list icons

	Time to first fixation		Total fixation duration		Fixation count	
	Western	Arab	Western	Arab	Western	Arab
Mean	7.85 s	5.69 s	0.47 s	1.15 s	1.8	4.4
Df	18		18		18	18
t Stat	0.9371		−2.2969		−2.0413	
P(T <= t) one-tail	0.1805		0.0169		0.0280	
Observations	10	10	10	10	10	10

The fixation count is the number of times a participant fixates on the wish list icon. The mean fixation count of 4.4 s for the wish list icon in the Arab design was significantly greater than that of the mean fixation count of 1.8 s on the alternate design (t-value of −2.04135; p-value = 0.0280). Once again the culturally adapted icon scored higher by attracting more number of fixations than the Western-influenced design.

Based on analyzing the questionnaire, we found that most participants felt the icons on the Western-influenced site were "international" or "regular." The participants related most to the Arab-influenced website and felt that it reflected their culture. Most of the participants, i.e., eight of 10, retrospectively referred to the wish list and other culturally attuned icons and felt that they helped them relate the website to Arab culture. For instance, one of the participants said, "The icons for men and women were interesting, I thought, because of the cultural aspect, like they were dressed culturally appropriate. And I thought the genie lamp was interesting for the wish list." A majority of the participants, nine of 10, felt that the cultural markers on the website represented their culture. Analysis of the responses to the question on perceived cultural affinity showed that the Arab participants felt more cultural affinity for the Arab-influenced website (t Stat = 3.641115; p (T <= t) two-tail 1.05E−07), further confirming the qualitative data obtained from the interviews.

4 Conclusions and Implications

The positive influence of culture on the design of websites and related consumer acceptance has been studied and documented well. As described in the paper, we created numerous visual metaphors based on themes from Arabic folklore. Of these we selected a few for inclusion in the design of a website for an eye-tracking experiment. The results showed that cultural affinity appeared to be higher for the Arab-influenced design, and greater attention was paid to the culturally attuned wish list icon. Some major limitations of the study design included the use of English language to drive the content and the small sample size of the participant pool.

As prior studies have mentioned, the use of cultural elements and cultural affinity affect the way users view websites. This research provides interesting insights for businesses, Web designers, and other researchers in Arab countries. Businesses can use the findings of this research to improve their websites and better reach their target audiences. The findings of this research will also help Web designers communicate better with users. This study may encourage other researchers to further investigate culturally attuned visual metaphors among other cultural elements in e-commerce websites.

Acknowledgments. This work was partly made possible by NPRP grant 5-1393-6-044 from the Qatar National Research Fund (a member of the Qatar Foundation). The statements made herein are solely the responsibility of the authors.

References

1. Marcus, A., Gould, E.W.: Cultural dimensions and global web user-interface design: What? So what? Now what? In: Proceedings of the 6th Conference on Human Factors and the Web, Austin, Texas, USA (2000)
2. Robbins, S.S., Stylianou, A.C.: A study of cultural differences in global corporate websites. J. Comput. Inf. Syst. **42**, 3–9 (2002)
3. Singh, N.: Culture and the world wide web: a cross-cultural analysis of web sites from France, Germany, and USA. Am. Mark. Assoc. **14**, 30–31 (2003)
4. Khanum, M.A., Fatima, S., Chaurasia, M.A.: Arabic interface analysis based on cultural markers. Int. J. Comput. Sci. Issues **9**, 255–262 (2012)
5. Lakoff, G., Johnson. M.: Metaphors We Live By. University of Chicago Press, Chicago (1980)
6. Forceville, C.: Pictorial metaphor in advertisements. Metaphor Symbolic Act. **9**, 1–29 (1994)
7. Shen, S.T., Woolley, M., Prior, S.: Towards culture-centered design. Interact. Comput. **18**, 820–852 (2006). Chicago
8. Heukelman, D., Obono, S.E.: Exploring the African village metaphor for computer user interface icons. In: Proceedings of the 2009 Annual Research Conference of the South African Institute of Computer Scientists and Information Technologists, pp. 132–140. ACM (2009)
9. Liginlal, D., Rushdi, M., Meeds, R., Ahmad, R.: Localization for a high context culture: an exploratory study of cultural markers and metaphors in Arabic ecommerce websites. In: Proceedings of the 2014 International Conference on E-Commerce, E-Business, and E-Service, pp. 21–28. Guilford Press/Taylor & Francis, London (2014)
10. Poole, A., Ball, L.J.: Eye-tracking in human-computer interaction and usability research. In: Ghaoui, C. (ed.) Encyclopedia of Human Computer Interaction, Idea Group Reference, vol. 1, pp. 211–219 (2006)

Creditworthiness Analysis in E-Financing Businesses - A Cross-Business Approach

Kun Liang[1,2(✉)], Zhangxi Lin[2], Zelin Jia[2], Cuiqing Jiang[1], and Jiangtao Qiu[2,3]

[1] Shcool of Management, Hefei University of Technology,
193 Tunxi Road, Hefei 230009, China
{liangkun_fd,jiangcuiq}@163.com
[2] Jerry S. Rawls College of Business Administration, Texas Tech University,
703 Flint Ave, Lubbock, TX 79409, USA
{zhangxi.lin,zelin.jia}@ttu.edu
[3] School of Economic Information Engineering,
Southwestern University of Finance and Economics, 55 Guanghuacun Road,
Chengdu 610074, China
qjt163@163.com

Abstract. To cope with the challenge of data scarcity in creditworthiness analysis for e-financing business, this paper proposes a cross-business analysis approach based on the assumption of behavior consistency for client in different e-commerce environments. By this approach we can analyze individuals' creditworthiness by associating financial data on lending platforms and cross-business non-financial data on social media. We conceived three creditworthiness assessment models, and conduct the experimental study on Ant Financial Co-Creation Data Platform. The results verify that our cross-business creditworthiness analysis approach is effective.

Keywords: Online lending · Creditworthiness · Cross-business data · Modeling · Data mining

1 Introduction

Following the trend of financial disintermediation, innovative e-financing businesses, such as P2P (Peer-to-Peer), P2B (Peer-to-Business), P2G (Private-to-Government), crowdfunding, and so on, provide diversified funding services directly to small businesses and consumers through various online platforms. For example, Alibaba launched its online microloan services in 2010, namely Aliloan, which has issued about 25 billion dollars loans by 2014, benefiting more than one million small and micro-sized enterprises [1]. In review an online loan application, a client's ability and willingness to fulfill contracts, i.e. his/her creditworthiness, is an important indicator [2], because creditworthiness analysis can effectively reduce the information asymmetry between financial suppliers and borrowers in the e-financing platform, and improve the accuracy of loan decisions.

Existing creditworthiness analysis methods predicted individuals' creditworthiness mainly based on the historical data formed in the lending business, such as payment

© Springer International Publishing Switzerland 2016
V. Sugumaran et al. (Eds.): WEB 2015, LNBIP 258, pp. 181–185, 2016.
DOI: 10.1007/978-3-319-45408-5_18

history, credit usage, length of credit history; most of them are transactional financial data [3, 4]. However, this kind of data is hard to obtain in e-financing businesses [3], while user-generated data are abundant but of non-financial, such as the social networking data in social media platforms and online reputation scores in electronic markets. These cross-business non-financial data can effectively reflect individuals' creditworthiness from multiple perspectives [4]. For example, one person can play different roles in various e-commerce businesses. On one hand, he or she can be a borrower in an online lending business; on the other hand, he or she can also be a seller in a C2C market. In this context, many cross-business data, such as his or her online reputation scores formed in trading businesses (C2C transactions), can be used to assess his or her creditworthiness in the lending business [4].

This paper is intended to investigate that how individuals' creditworthiness-related non-financial data can be used in the e-financing business. Specifically, we study the following two questions. First, how effective the cross-business non-financial data can be used to analyze individuals' creditworthiness in the e-financing business? Second, how to select reasonable cross-business non-financial indicators for creditworthiness analysis in e-financing businesses?

The contribution of this study can be summarized as following: (1) We proposed a cross-business creditworthiness analysis approach to overcome the data scarcity problem of creditworthiness analysis in e-financing businesses; (2) We revealed that the social capital theory can be helpful in selecting reasonable cross-business non-financial data to improve the performance of creditworthiness analysis for e-financing businesses.

2   A Cross-Business Creditworthiness Analysis Approach

In this section, we propose a cross-business creditworthiness analysis approach based on the consistent attribute of creditworthiness. Adelson et al. (2009) proposed that creditworthiness can be understood as the relative ranking of default frequency, and this relative ranking is consistent in different business and scenarios [5]. In addition, individuals' creditworthiness is their ability and willingness to fulfill contract. Although individuals may play different roles and have to fulfill different contracts in various businesses, their creditworthiness is influenced by some common factors, such as their morality and sense of responsibility. Therefore, one person's creditworthiness is consistent in different businesses.

We select individuals who are borrowers in the online lending market, and also sellers in the C2C market. Assuming that an individual's creditworthiness is consistent across different online platforms [5], we analyze these individuals' creditworthiness in lending business by considering their behaviors in online commodity transactions, which could be reflected by their reputation and social capital. An individual's default behaviors could destroy his/her reputation that was previously accumulated as the intangible asset [6]. From this sense, the online reputation can effectively reflect one's creditworthiness. Furthermore, maintaining good creditworthiness can help one to gain more social resource such as friends, job opportunity, and so on [4]. Based on this idea, we conceive three models to analyze individuals' creditworthiness in e-financing businesses with different indicator sets (see Fig. 1). Model 1 analyzes individuals' creditworthiness based on the indicator set which only contain financial factors (Fa); Model 2 counts online reputation factors (Fb) into the indicator set; Model 3 further adds social capital factors (Fc) into the indicator set. We are to study whether we can analyze individuals' creditworthiness in e-financing business more accuracy by adding cross-business non-financial factors (Fb and Fc).

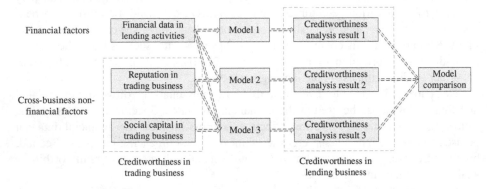

Fig. 1. A cross-business creditworthiness analysis approach

3   Experimental Study

We conduct an experimental study in Ant Financial Co-Creation Data Platform, which became available in June 2015 complying with the agreement between Ant Financial and Tongji University, to validate the feasible of proposed cross-business

Table 1. The performance of different creditworthiness assessment models

Models	FN	TN	FP	TP	TPR	TNR	Cost
DT 1	321	1293	161	205	0.3897	0.8893	3.7723
DT 2	250	1203	251	276	0.5247	0.8274	3.0243
DT 3	239	1173	281	287	0.5456	0.8067	2.9195
NN 1	320	1331	123	206	0.3916	0.9154	3.7348
NN 2	295	1328	126	231	0.4392	0.9133	3.4517
NN 3	283	1301	153	243	0.4620	0.8948	3.3334
LR 1	345	1336	118	181	0.3441	0.9188	4.0165
LR 2	302	1328	126	224	0.4259	0.9133	3.5315
LR 3	291	1329	125	235	0.4468	0.9140	3.4054
SVM 1	421	1401	53	105	0.1996	0.9635	4.8387
SVM 2	363	1369	85	163	0.3099	0.9415	4.1991
SVM 3	340	1351	103	186	0.3536	0.9292	3.9492

Model n (include DT n, NN n, LR n, and SVM n), n = 1,2,3.
Model 1, Model 2, and Model 3 refer to the models that use the indicator sets of Fa, Fa + Fb, Fa + Fb + Fc, respectively. For example, DT 1 is the decision tree model which adopt only financial indicators (Fa)

creditworthiness analysis approach. We compiled a dataset of 6,598 observations containing 366 variables dated 20 April, 2015. Each observation represents an individual's records in both online lending market and C2C market. The target variable, i.e. the creditworthiness of an individual is set either good or bad, determined by whether there is a late payment record in his/her loan history. We select four classification technologies to construct our creditworthiness assessment model, i.e. Logistic Regression (LR), Decision Tree (DT), Support Vector Machine (SVM), and the Neural Network (NN). We use three criteria to evaluate the performance of each model with different indicator sets. They are TPR, TNR and Cost, which were defined in [7]. TPR is the percentage of correctly classified bad creditworthiness. TNR is the percentage of correctly classified good creditworthiness. Cost is a aggregative criteria consider both TPR and TNR. The results for the comparison of different models and indicator sets are summarized in Table 1.

In Table 1, the TPR of Model 2 is higher than that of Model 1. This result indicates that the model can identify more bad creditworthiness individuals when add the reputation factors. However, the TNR of Model 2 is slightly less than that of Model 1. This result shows that the ability to recognize the good creditworthiness individuals has a little decrease when add the reputation factors. Due to the different misclassification cost of good and bad creditworthiness [7], we use the criterion of Cost to determine the Models' comprehensive ability to differentiate the good and bad creditworthiness. In general, Model that has a lower Cost is better in creditworthiness analysis. In Table 1, the Cost of Model 1 is higher than that of Model 2, which indicates that individuals' reputation in Alibaba C2C business has significant prediction ability to their creditworthiness in Ali-loan business. Similarly, the Cost of Model 2 is higher than that of Model 3, which indicates that individuals' social capital in Sina micro-blog (reflected the social capital accumulated in C2C transaction) has significant prediction ability to their creditworthiness in Ali-loan business. Compared to the performance of Model 1–3, we revealed that individuals' creditworthiness in an e-financing business can be predicted by their creditworthiness in the trading business which is analyzed by their reputation and social capital (cross-business non-financial factors) formed in C2C transaction activities. This result verifies the consistent attribute of creditworthiness [5]. Now, we can answer the two questions proposed in the Introduction. First, we can analyze individuals' creditworthiness in the e-financing business based on the cross-business non-financial data. Second, we can select reasonable cross-business non-financial indicators according to the reputation and social capital theory.

4   Conclusions

In order to solve the data scarcity problem in creditworthiness analysis for the e-financing business, this study proposed a cross-business creditworthiness analysis approach according to the consistent attribute of creditworthiness. This paper makes several contributions to the literature. We proposed a new creditworthiness analysis approach for e-commerce businesses, which analyze individuals' creditworthiness in one business by considering their creditworthiness in the other business. In addition, we enlarge the application scope of reputation and social capital theory to cross-business creditworthiness analysis. From the perspective of practices, we provide a better creditworthiness assessment model for e-financing businesses, which can effectively improve the accuracy of loan decisions. The methodology of cross-business creditworthiness analysis can also be used in other e-commerce businesses, such as P2P lending. From this sense, we could provide e-commerce platforms with more insights about individuals' creditworthiness from various business perspectives.

However, this research is limited and must be further expanded. In this paper, evaluation indicators are restricted to structured data, and there is a lack of relevant indicators to reflect the relational aspect of social capital in creditworthiness analysis. In next step, we will analyze the creditworthiness through the relationship types of social network ties and indirect social network ties, which can effectively reflect the effect of relational aspect of social capital on creditworthiness.

References

1. The actual financing cost is only 6.7 %: Ali finance shocks the private lending, http://business.sohu.com/20130320/n369488586.shtml
2. Safi, R., Lin, Z.: Using non-financial data to assess the creditworthiness of businesses in online trade. In: PACIS Proceedings (2014)
3. Wang, Y., Li, S., Lin, Z.: Revealing key non-financial factors for online credit-scoring in e-financing. In: 2013 10th International Conference on Service Systems and Service Management (ICSSSM), pp. 547–552. IEEE, July 2013
4. Lin, Z., Whinston, A.B., Fan, S.: Harnessing internet finance with innovative cyber credit management. Finan. Innov. 1(1), 1–24 (2015)
5. Adelson, M., Ravimohan, R., Griep, C., Jacob, D., Coughlin, P., Bukspan, N., Wyss, D.: Understanding standard & poor's rating definitions. Standard & Poor's (2009). http://www.standardandpoors.com/spf/delivery/assets/files/Understanding_Rating_Definitions.pdf
6. Van den Bogaerd, M., Aerts, W.: Does media reputation affect properties of accounts payable? Eur. Manage. J. 33(1), 19–29 (2015)
7. Lessmann, S., Baesens, B., Seow, H.V., Thomas, L.C.: Benchmarking state-of-the-art classification algorithms for credit scoring: a ten-year update. Eur. J. Oper. Res. 124–136 (2015)

Empirical Investigation of Partnership and Mediating Effect of Mode of Partnership on Innovation Outcome of IT Firms

Ashish Kumar Jha[✉] and Indranil Bose

Indian Institute of Management Calcutta, Kolkata, India
{ashishkj11,bose}@iimcal.ac.in

Abstract. Higher cooperation with various players is widely believed to be an important determinant of innovation productivity. The players include suppliers, clients, competitors, universities etc. However the importance of these probable partners and their differential impact on product innovation vis-à-vis process innovation is an issue which has not been explored in extant literature. Also another important consideration is whether having cooperation partners directly impacts the innovation productivity or the relationship is mediated by the choice of mode of partnership. While the extant literature has established inter-firm cooperation to be important for innovation at a generic level, we find that it is the dynamics of the partnership characterized by the cooperation partner and more importantly the mode of the partnership which determines the success of such a cooperative effort.

Keywords: Product innovation · Process innovation · Technological innovation in IT firms · Inter-firm cooperation

1 Introduction

One of the reasons why the study of firm size's impact on innovations has become outmoded is due to the impact of blurring of line differentiating the differences between the firms [1, 2]. Firms are more and more interconnected and networked. The efforts for innovation are not being put in by one firm by generally by a lot more players. These may include either the other players in the value chain or other players in an industry. Various researches have established the importance of cooperation as important for innovation in some form or other [2, 3]. However the dynamics of cooperation for innovation still remains one of the less understood areas of innovation studies.

The important questions in the dynamics of collaborative/cooperative innovation efforts pertain to how the collaboration actually takes place. More than the eventual decision of partnership with a given external entity, how the partnership is executed, can determine the outcome of such an effort. Based on the data of IT firms of Europe, we attempt to answer the questions related to these dynamics and extend the theoretical underpinnings of innovation cooperation.

© Springer International Publishing Switzerland 2016
V. Sugumaran et al. (Eds.): WEB 2015, LNBIP 258, pp. 186–190, 2016.
DOI: 10.1007/978-3-319-45408-5_19

More specifically, our current work aims to shed more light on the dynamics of cooperation by evaluating the importance of the cooperation partners and how these partnerships impact the innovation productivity. The critical questions that we aim to answer through this work are

(a) Is presence of cooperation partner significant for increasing innovation productivity for product innovation and process innovation?
(b) Is the impact of partner on innovation productivity mediated by the mode of such a partnership?

2 Theory

Firm level cooperation is increasingly been considered as an important determinant in firm's survival and success in todays networked economies [4]. Innovation is no exception to this trend and cooperative firms have been found to have higher R&D intensity [5]. Increase in R&D intensity is not the only determinant leading to higher amount of cooperation amongst firms. Belderbos et al. [6] found in their work that higher cooperation increases firms' profitability from R&D activities. R&D activities, as is commonly known, are very resource intensive and cooperation amongst firms helps in sharing resources, investments along with knowledge thus increasing the success of these efforts while considerably reducing costs for the firms. One more aspect in the increasing clamor for cooperation in innovation is the influx of new perspectives to cater to higher demand for innovation from market. Various firms have consistently increased their R&D investments to create innovation units to develop new products for their customers and clients and external input in such efforts increases the outcome significantly. These factors also elucidate why the high technology intensive industries have a higher concentration of cooperative innovation [7].

Although cooperation has been established to be an important factor for innovation productivity, its differential impact on product and process innovation is yet to be studied. A deeper understanding implications of cooperation on both product and process innovation is necessary to better plan the cooperation efforts of firms.

The aim of the study is to establish the presence of partner as a significant contributor to the innovation outcome and to further establish the mediation effect of mode of

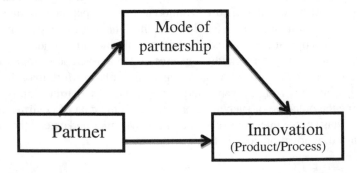

Fig. 1. Mediated research model of cooperation impacting innovation productivity

partnership in this if any. We base our model on the prior such works for mediation analysis [8, 9]. Figure 1 shows the mediated research model which would be used in this study. The mediation effect is validated through Sobel test in our model using the three step model approach [8, 9].

2.1 Data

The analysis in this study is being performed on European IT firms' data. The data has been drawn from the World Bank Enterprise Survey's innovation dataset. We have used data for year 2012 which is the latest year for which the data is available. The innovation survey, which is conducted by the World Bank as a part of the Enterprise Survey, is designed using the guidelines that are provided in the Oslo manual for innovation [10]. These surveys are conducted physically by World Bank representatives and are made available for use by researchers around the world on request. Table 1 presents some of the descriptive statistics of the dataset.

Table 1. Descriptive statistics of properties of the firms.

Variable	Mean	Std. Dev.	Min.	Max
Firm revenue[#]	65.3	271	0.01	2000
Average annual spend on intramural R&D[#]	1.1	2.08	0.007	7
Firm Age	17.53	11.63	4	70
Firm Employees	47.18	87.82	2	600

[#] All values are in million US$

We use logistic regression to analyze the research model presented in the previous section. The dependent variables in the different models are product innovation outcome, Mode of partnership and process innovation outcome. All of these are binary variables leading to our choice of the regression method. The independent variables used in the study are revenue, age, degree of competition, use of specialized software like ERP, presence of foreign collaboration and contribution of main product in overall sales of firm.

3   Preliminary Results

This is a research in progress paper and most analysis is still being conducted. However some preliminary analyses have been conducted which sheds some light on the research questions. The results show that 'presence of partner' is a significant variable in explaining the innovation outcome for both product and process innovation. The question of relative significance of different partner variables would be further explored in this work. With respect to age of firm we found that age is a significant predictor for product innovation, it is insignificant for process innovations. Our results indicate that

IT firms tend to be less productive in terms of product innovation as they age. These results are in line with prior such studies [11, 12].

Further as we also found that use of specialized software like ERP and SCM systems are significant for product innovation and not process innovations. This shows that process innovations are extremely internal to the firm and there is a limited amount of impact that external partners have on process innovation. However close coordination with different partners and closer maintenance of firm level information may promote higher product innovation. Another interesting insight is that presence of a foreign partner in any form promotes process innovation in a firm. This result is very intuitive as firms might have to revamp their internal process to adapt the firms' processes to other processes followed at the partner firms abroad while such a partnership may not necessarily lead to higher product innovation within these firms.

4   Future Work and Conclusion

The paper presents a novel view of the innovation dynamics in IT industry. Using an extremely rich, reliable but infrequently used Enterprise Survey dataset, we aim to be able to present interesting insights on the innovation process. The preliminary results indicate that innovation partners are significant players in the innovation process of firms, for both product as well as process innovations.

A greater amount of work on generalizability would be required to be done using datasets of multiple geographies and firms. The results also would be improved by further drilling into the comparative estimates of the benefits of different partners for different innovation outcome, which we are in process of analyzing.

A lot more work needs to be done to further understand the complex and intricate relationship between cooperation partner and focal firm. Our current work is a step in the direction of unearthing some aspects of this relationship and its implications. The most important contribution of our current research is developing the understanding that collaborating closely with partner is more important for the focal firm if the agenda for innovation efforts is product innovation. However, licensing and patent transfer work better for the focal firm for process innovations. Also while having an innovation partner is important for process innovation, the importance is much more for product innovation.

Acknowledgments. The authors like to thank the World Bank Enterprise survey for providing access to firm level micro data for this research to be completed (Enterprise Survey). The authors like to state that all inferences drawn in the paper and all views presented are authors' own and are in no way representative of World Bank's views or inferences.

References

1. Teece, J.D.: Competition, cooperation, and innovation: organizational arrangements for regimes of rapid technological progress. J. Econ. Behav. Organ. **18**(1), 1–25 (1992)
2. Hagedoorn, J.: Inter-firm R&D partnerships: an overview of major trends and patterns since 1960. Res. Policy **31**(4), 477–492 (2002)

3. Hagedoorn, J.: Understanding the rationale for strategic technology partnering: inter-organizational modes of cooperation and sectoral differences. Strateg. Manag. J. **14**(5), 371–385 (1993)

4. Abramovsky, L., Kremp, E., López, A., Schmidt, T., Simpson, H.: Understanding cooperative innovative activity: evidence from four European countries. Econ. Innov. New Tech. **18**(3), 243-265 (2009)

5. Sampson, R.: R&D Alliances and firm performance: the impact of technological diversity and alliance organization on innovation. Acad. Manage. J. **50**(2), 364–386 (2007)

6. Belderbos, R., Carree, M., Lokshin, B.: Complementarity in R&D cooperation strategies. Rev. Ind. Organ. **28**(4), 401–426 (2006)

7. Miotti, L., Sachwald, F.: Cooperative R&D: why and with whom? An integrated framework of analysis. Res. Policy **32**(8), 1481–1499 (2003)

8. Baron, R.M., Kenny, D.A.: The moderator-mediator variable distinction in social psychological research: conceptual, strategic, and statistical considerations. J. Pers. Soc. Psychol. **51**(6), 1173–1182 (1986)

9. Aggarwal, R., Singh, H.: Differential influence of blogs across different stages of decision making: the case of venture capitalists. MIS Quart. **37**(4), 1093–1112 (2013)

10. OECD: Proposed guidelines for collecting and interpreting technological innovation data. In: The Measurement of Scientific and Technological Activities: Oslo Manual. OECD Publishing (1997)

11. Hansen, J.A.: Innovation, firm size, and firm age. Small Bus. Econ. **4**(1), 37–44 (1992)

12. Stuart, T.E.: Interorganizational alliances and the performance of firms: a study of growth and innovation rates in a high-technology industry. Strateg. Manage. J. **21**(8), 791–811 (2000)

Privacy Calculus Theory and Its Applicability for Emerging Technologies

Adrija Majumdar[(✉)] and Indranil Bose

Management Information Systems, Indian Institute of Management Calcutta,
Kolkata, India
{adrijam13,bose}@iimcal.ac.in

Abstract. One of the most important theories to be used and modified subsequently in the IS literature is the privacy calculus theory. This calculus governs the decision-making process of individuals to predict certain behavioral outcome like, disclosing personal information, intention to use an e-commerce site, in the presence of perceived privacy risk and perceived benefits. In this paper, we seek to analyze the relevancy of privacy calculus theory for certain emerging technologies and platforms (i) Bring Your Own Device (BYOD) (ii) Internet Of Things (IoT). We identify some of the perceived privacy risks and benefits in these emerging technologies. The insights gained from this study will enable researchers to further study the behavioral intention of organizations/people to use these technologies in the presence of privacy risks and benefits.

Keywords: Information privacy · Privacy calculus · Privacy risks · BYOD · IoT

1 Introduction

Privacy is not highlighted in terms of its physical, legal and behavioural aspects in the IS discipline. The focus of IS discipline is in on the information aspect of privacy. Often it is seen individuals indulge in a cost-benefit analysis in terms of whether to disclose their personal information or not. It has been found that consumers sacrifice a certain portion of their privacy in lieu of some benefits consisting of financial incentives or convenience [1]. A growing amount of literature has emphasized on the privacy-related decision making as a cognitive process by which individuals weigh the (a) anticipated costs or the risks of disclosing information (b) perceived benefits from disclosing such information. In this research in progress, we focus on the use of such a "privacy calculus" in information privacy literature. To the best of our knowledge, this is the first paper in IS literature which critically examines the academic literature on privacy calculus to develop a greater understanding on its applicability for emerging technologies. Insights shared in this study will help academic scholars to appreciate the privacy calculus better in these emerging fields.

© Springer International Publishing Switzerland 2016
V. Sugumaran et al. (Eds.): WEB 2015, LNBIP 258, pp. 191–195, 2016.
DOI: 10.1007/978-3-319-45408-5_20

2   Theoretical Background

The term privacy calculus was first used to denote the "calculus of human behaviour" [2]. This calculus governs the decision-making process of individuals to decide whether to disclose personal information. Privacy calculus acknowledges the contribution of expectancy theory that proposes that human agents act in ways that maximizes the positive outcomes and minimizes the negative results. The principal components that are connected to the concept of privacy calculus are perceived risk and perceived benefits. In this section, we briefly discuss the notion of 'Privacy Risk' and 'Privacy Benefits'.

2.1   Privacy Risk

Perceived risk is the fear that the consumer's private information could be used by organizations for unfair purposes, like price discrimination or it could be sold to other third parties who should not have access. It is the subjective evaluation of potential privacy related losses that could affect a consumer. Privacy risks can be categorized along five risk dimensions i.e. social, financial, time, psychological and physical [3]. Existing literature shows a significant positive relationship between privacy concerns and the risk beliefs [4]. These privacy concerns are in turn dependent on the way the personal information is collected, on the extent of control an individual has over personal information and on his/her awareness of information practices. There is reasonable accord among IS scholars that perceived privacy risks have a negative impact on the intention to disclose personal information [4]

2.2   Perceived Benefit

It is the subjective evaluation of potential gains or positives. IS scholars have identified three significant factors that contribute to the perceived benefit component, namely, the inclination of consumers towards financial rewards, the love for personalization and the desire for social adjustment benefits [5]. Reducing search time for locating appropriate promotional messages, providing convenience for instant access to personalised messages, having an overall satisfaction with the personalised service are some of the items used in the literature to capture the notion of perceived benefit due to personalisation. On one hand, consumers treasure the benefits of personalized guidance and, on the other hand, they have privacy concerns about providing personal information that are required to build these features of personalization.

3   Use of Privacy Calculus in Emerging Technologies

In this section we analyze the applicability of privacy calculus theory in the context of two emerging trends in Information Technology, (i) BYOD and, (ii) IoT. According to the industry reports, it is expected that around 78.48 % of organizations in the USA will have BYOD activity by 2018 [6]. According to International Data Corporation, the worldwide IoT market is estimated to grow at a CAGR of 16.9 % and touch $1.7 trillion in 2020 [7]. IoT aids real time decision making and handles challenges with aging workforce. IoT promises to create billion dollar markets in the form of smart cities, smart factories, smart supply chains etc.

3.1 Bring Your Own Device (BYOD)

Mobile devices offer immense convenience, efficiency and flexibility for users [8]. BYOD is defined as the use of privately owned mobile devices for official corporate work [9]. Employees could use their privately owned mobile devices for various reasons, viz. to access the official mail, to create and store and manage official data, to access company databases. The usage of BYOD, however, entails security and privacy concerns. The connection of privately owned devices with the corporate facilities increases the chance of malware intrusion and the likelihood of data loss and theft [8]. Employees' privacy concerns could be heightened as they fear their loss of private information into the hands of their employers. Furthermore, owing to advanced GPS technology, the location data of the employees could be tracked in real time which aggravates their privacy concerns. However, the use of BYOD also provides sufficient advantage for both the organization and the employees. It increases the job satisfaction of the employees as it increases their personal freedom, jobholders can now work according to their preferred place and time. Thus, BYOD is associated with a set of contrary factors and it provides an appropriate background for the use of Privacy calculus theory. In fact, researchers have integrated the privacy calculus framework and the technology acceptance model to study the influence of security, privacy and legal concerns on the intention to use BYOD mobile devices [8]. Figure 1 lists some of the important decision facilitators and inhibitors of this domain.

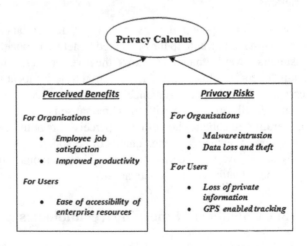

Fig. 1. Privacy calculus framework for BYOD

3.2 Internet of Things (IoT)

Internet of Things comprises of sensor-based IS services, whose functionality is facilitated by the identification technologies consisting of barcodes, RFID, global satellite communication, etc. [10]. There is a huge opportunity of collecting real-time data from these new technological artifacts. These data could be mined and organizations could earn a competitive advantage. IOT is gaining immense popularity in the wireless tele-communications domain. The primary essence of IOT is its pervasive presence and interaction and co-operation with their neighboring sensors to reach commonly shared goals [11]. The advantage gained from implementing such technologies is immense; however it also entails sacrificing a share of consumer's privacy. For example, a typical IOT technology RFID provides benefits especially for consumers in the logistics and marketing domain for tracking purpose. However, there is sufficient privacy risks associated with this technology. RFIDs could suffer from the clandestine scanning of tags by unauthorized personnel. Furthermore, real-time tracking of location could be over intrusive at times. Thus, we notice a set of trade-offs exists in this emerging technology that provides us the background to use the theory of privacy calculus. Figure 2 highlights some of the important decision facilitators and inhibitors for IoT.

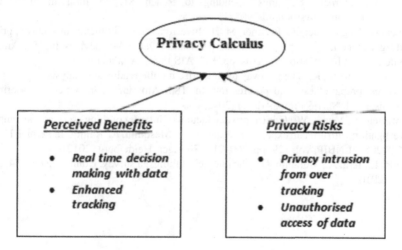

Fig. 2. Privacy calculus framework for IoT

4 Conclusion

This is a research in progress and we attempted to analyze the suitability of privacy calculus theory in the promising fields of BYOD and IoT. We plan to conduct interviews/focus group interviews with the stakeholders of these technologies to gain better insights on the perceived benefits and risks for these emerging fields. As a future work we want to additionally examine few other emerging areas like, cloud computing and online social networks with the lens of privacy calculus theory. The insights gained from this study will help researchers to analyze the behavioral intention of organizations/people to use these technologies in the presence of privacy concerns.

References

1. Hann, I.H., Hui, K.L., Lee, S.Y.T., Png, I.P.: Overcoming online information privacy concerns: an information-processing theory approach. J. Manage. Inform. Syst. **24**, 13–42 (2007)
2. Laufer, R.S., Wolfe, M.: Privacy as a concept and a social issue: a multidimensional developmental theory. J. Soc. Issues. **33**, 22–42 (1977)
3. Youn, S.: Determinants of online privacy concern and its influence on privacy protection behaviors among young adolescents. J. Consum. Aff. **43**, 389–418 (2009)
4. Malhotra, N.K., Kim, S.S., Agarwal, J.: Internet users' information privacy concerns (IUIPC): the construct, the scale, and a causal model. Inform. Syst. Res. **15**, 336–355 (2004)
5. Smith, H.J., Dinev, T., Xu, H.: Information privacy research: an interdisciplinary review. MIS Quart. **35**, 989–1016 (2011)
6. Gandhi, V.: Bring Your Own Device (BYOD)—Key Trends and Considerations. Industry Report, Frost & Sullivan (2013)
7. Explosive Internet of Things Spending to Reach $1.7 Trillion in 2020, http://www.gartner.com/newsroom/id/2905717
8. Lebek, B., Degirmenci, K., Breitner, M. H.: Investigating the influence of security, privacy, and legal concerns on employees' intention to use BYOD mobile devices. In: 19th Americas Conference on Information Systems, pp. 1–8. AIS Press, Atlanta (2013)
9. Johnson, N., Joshi, K.: The pathway to enterprise mobile readiness: analysis of perceptions, pressures, preparedness, and progression. In: 18th Americas Conference on Information Systems, pp. 1–8. AIS Press, Atlanta (2012)
10. Kowatsch, T., Maass, W.: Critical privacy factors of internet of things services: an empirical investigation with domain experts. In: Rahman, H., Mesquita, A., Ramos, I., Pernici, B. (eds.) MCIS 2012. LNBIP, vol. 129, pp. 200–211. Springer, Heidelberg (2012)
11. Atzori, L., Iera, A., Morabito, G.: The internet of things: a survey. Comput. Netw. **54**, 2787–2805 (2010)

An IT Risk-Return Model to Study IT Governance of Indian Firms

Manas Tripathi[1], Arunabha Mukhopadhyay[1(✉)], and Indranil Bose[2]

[1] Indian Institute of Management Lucknow, Lucknow, India
{fpml3007,arunabha}@iiml.ac.in
[2] Indian Institute of Management Calcutta, Calcutta, India
bose@iimcal.ac.in

Abstract. In this paper, we measure IT governance using an IT risk-return model. We study impact of market and accounting return of Indian firms' vis-à-vis their IT investments for the period 2003–2014. We have focused on three sectors (manufacturing, IT, and banking) only due to paucity of data. We note IT investments make significant contribution to overall firm risk and lead to earning stabilization and reduction of firm risk. The impact of IT investment on IT return, measured, is positive and significant. However, IT return reduces, once we incorporate IT risk term in this model, which reflects the risk premium associated with gross IT return.

Keywords: IT risk · IT return · IT investment · IT value · IT capital · IT governance

1 Introduction

In 2014, the IT spending across the world was $3.8 trillion and is expected to grow by 2.4 % in 2015 [1]. NASSCOM reports, the revenue of the IT sector is close to US$150 billion in 2015, an increase of 13 % from the previous year [20]. However, the value and ROI seems a bit elusive [2]. Managers have been debating whether profitability of a firm increases due to investments in R & D, Advertising etc. or in IT [3, 9]. Nevertheless, there has been a widespread belief that investment in IT innovations will help to enhance products, services and customer intimacy for the firm [4]. IT investments are risky due to technological complexity and implementation challenges associated with them. McKinsey found 17 % of large IT projects go so badly that they can threaten the very existence of the company. On average, large IT projects run 45 % over budget, 7 %over time, while delivering 56 % less value than predicted [5]. Especially ERP systems have high failure rates [6–8].

Our aim is to study IT Governance using risk-return profile of IT investment for Indian firms. Here, we study how these new technologies are affecting risk-return profile for IT investment for Indian IT, BFSI and manufacturing sector. This study will also focus on risk adjusted IT return by incorporating IT risk into excess return.

There is a dearth of literature on the risk return relationship of IT investment vis-à-vis work on IT returns [10–13] and IT risk at the project-or systems level [14–18]. Researchers have developed proxy measure for IT risk and looked at risk-return profile

V. Sugumaran et al. (Eds.): WEB 2015, LNBIP 258, pp. 196–201, 2016.
DOI: 10.1007/978-3-319-45408-5_21

of IT investment for US firms [19].Our paper has six sections. The next section provides the theoretical background of our analysis. In Sect. 3, we describe our research framework and model. Section 4 explains about data. Section 5 covers results and conclusion is in Sect. 6.

2   Literature Review

ROA, ROI, ROE and ROS are common measures for capturing the rate of return from IT investments [10]. ROI of IT capital is reported to be about 81 % as compared to 6.26 % for non IT capital using a production function analysis [11–13, 21]. Market capitalization is a good measure of IT value [22]. Value from IT investments is generated through business process automation & transformation and effective decision making [23]. The value is reflected as increase in efficiency at process level [24] or firm level [11]. Indirect effect of IT on profitability has been studied by: Customer satisfaction [25], complementarity between computer investment and organizational investment [12], complementarities between IT resources and IT capabilities [26]. IT investments can enable firms to achieve both revenue growth and cost savings [27–30]. [19, 31] developed a proxy measure for IT risk and incorporate it into the IT return using production function analysis and market value specification. [32] study the impact of CRM investments on the stock price of firms.

3   Research Model

We will measure for overall firm risk (σ_{lt}) and then incorporate the IT risk (σ_{IT}) into it. The common measures of firm risk are standard deviation of (i) one-year daily stock returns following the investment(L_B_VOLA), (ii) realized annual earnings over 5 years following the investment(STD_PBDITA), [33]. Our model proposes that investment in IT capital(ITA), non-IT capital (K_Asset), expenses on R&D (RD), advertisement (AD), leverage (DE), and size(SZ) of the firm may have the impact on overall firm risk [19].

RQ1: How risky are IT investments vis-à-vis other types of capital investments?

$$Model\ 1 : \sigma_{lt} = \sigma_{IT} * IT\ Asset_{lt} + \sigma_K * K\_Asset_{lt} + \beta_1 * Size_{lt}$$
$$+ \beta_2 * Leverage_{lt} + \beta_3 * RD_{lt} + \beta_4 * AD_{lt} + \varepsilon_{lt} \tag{1}$$

RQ2: Does IT risk vary across industry segments (BFSI, IT and Mftg)?

$$Model\ 2 : \sigma_{lt} = \sum_j \sigma_{IT},j + \sigma_{IT} * IT\ Asset_{lt} + \sigma_K * K\_Asset_{lt}$$
$$+ \beta_1 * Size_{lt} + \beta_2 * Leverage_{lt} + \beta_3 * RD_{lt} + \beta_4 * AD_{lt} + \varepsilon_{lt} \tag{2}$$

Overall firm return (TV) depends on investment in IT capital stock (ITA), non-IT capital stock (KA), other asset (OA) and expenses like R&D (RD) and advertisement (AD) [12].

RQ3: Does IT return impact overall firm return? $\{\beta_1$: IT return coefficient$\}$.

$$Model\ 3 : TV_{lt} = \beta_0 + \beta_1 * IT_{lt} + \beta_2 * K_{lt} + \beta_3 * OA_{lt} + \beta_4 * RD_{lt} + \beta_5 * AD_{lt} + \varepsilon_{lt} \qquad (3)$$

RQ4: Does IT risk (σ_{IT}) have impact return on IT investment and productivity and market value(TV) of firms? [12, 19]. $\{\beta_6$: IT risk coefficient at industry level$\}$.

$$Model\ 4 : TV_{lt} = \beta_0 + \beta_1 * IT_{lt} + \beta_{2*}K_{lt} + \beta_3 * OA_{lt} + \beta_4 * RD_{lt} + \beta_5 * AD_{lt} + \beta_6 * \sigma_{ITj} + \varepsilon_{lt} \qquad (4)$$

4 Data

We have used financial and accounting data of Centre for Monitoring Indian Economy (CMIE) Prowess database for the period from April 2005 to March 2014 for this study. The final sample dataset consists of 135 companies taken from CNX-500 database.

5 Results

Table 1 shows coefficient of ITA is negative and insignificant, which indicates they are not risky for Indian firms as compared to other type of capital investments. Table 2 shows coefficient of ITA is significant at 5 % level and negative, which indicates that they are associated with reduced firm risk [12]. Table 3 illustrates IT risk coefficients for BFSI and IT sectors are insignificant. While for manufacturing sector, it is significant and has a negative coefficient, indicating reduced firm risk with IT investments.

Table 1. Pooled IT risk (L_B_VOLA)

Dep var: L_B_VOLA

Var	Cof	SE	t	p
AD	0.0	0.01	2	.06*
DE	−0.0	0.00	−3	.00#
ITA	−0.7	0.97	−1	.48
ITA1	−	−	−	−
ITA2	−	−	−	−
ITA3	−	−	−	−
KA	−0.7	0.21	−3	.00#
RD	−0.4	0.77	−1	.60
SZ@	−0.2	0.02	−10	.00#
R^2	0.4			
AdjR²	0.3			

#,* denote significance at 1 % and 10 % level respectively. @size is adjusted for weighted least square

Table 2. Pooled IT risk (STD_PBDITA)

Dep var:: STD_PBDITA

Var	Cof	SE	t	p
AD	0.1	0.06	2	0.02$
DE	−0.0	0.00	−1	0.24
ITA	−0.1	0.07	−2	0.04$
ITA1	−	−	−	−
ITA2	−	−	−	−
ITA3	−	−	−	−
KA	0.0	0.02	1	0.25
RD	1.2	0.14	8	0.00#
SZ@	0.0	0.00	1	0.37
R^2	0.2			
AdjR²	0.1			

#,$ denote significance at 1 % and 5 % level respectively

Table 3. Industry - IT risk

Dep var:: L_B_VOLA

Var	Cof	SE	t	p
AD	0.0	0.01	2	0.10*
DE	−0.0	0.00	−4	0.00#
ITA	−	−	−	−
ITA1	7.3	8.77	1	0.41
ITA2	−0.7	0.97	−1	0.49
ITA3	−17	6.90	−3	0.02#
KA	−0.7	0.21	−3	0.00#
RD	−0.6	0.78	−1	0.44
SZ@	−0.2	0.02	−10	0.00#
R^2	0.4			
AdjR²	0.3			

#,* denote significance at 1 % and 10 % level respectively

Table 4. IT return: market value specification

Dependent variable: TV

Variable	Cof	SE	t	p
ITA	14.7	5.26	2.80	0.00#
RD	155.9	8.46	18.42	0.00#
OA	0.4	0.02	22.02	0.00#
NON_ITA	1.8	0.32	5.79	0.00#
AD	31.8	2.10	15.15	0.00#
RISK	–	–	–	–
R^2	0.6			
Adj R^2	0.5			

#denote significance at 1 % level

Table 5. Risk adjusted IT return

Dependent variable: TV

Variable	Cof	SE	t	p
ITA	13.8	5.3	2.6	0.01#
RD	155.4	8.4	18.4	0.00#
OA	0.4	0.0	20.5	0.00#
NON_ITA	1.9	0.3	5.8	0.00#
AD	32.1	2.1	15.1	0.00#
RISK	784.6	563.0	1.4	0.16
R^2	0.6			
Adj R^2	0.5			

#denote significance at 1 % level

Table 4 shows that the coefficient for ITA is 14.7, which represents the contribution of IT asset towards the overall firm value [12, 19]. Table 5 shows the coefficient of ITA reduces to 13.8 and is consistent with the analysis of [19]. This magnitude of reduction reflects the risk premium associated with gross IT return.

6 Conclusion

From our IT Risk-return model, we can infer that IT governance for manufacturing sector is better as compared to IT and BFSI. Our model also notes the risk premium is associated with gross IT return.

References

1. Stamford, C.: Gartner says worldwide IT spending on pace to reach \$3.8 trillion in 2014, Gartner, January 2014
2. Westerman, G., Hunter, R.: IT Risk: Turning Business Threats into Competitive Advantage. Harvard Business School Press, Boston (2007)
3. Dewan, S., Kraemer, K.L.: Information technology and productivity: evidence from country-level data. Manag. Sci. **46**(4), 548–562 (2000)
4. Swanson, E.B.: Information systems innovation among organizations. Manag. Sci. **40**(9), 1069–1092 (1994)
5. Bloch, M., Blumberg, S., Laartz J.: Delivering large-scale IT projects on time, on budget, and on value, McKinsey, October 2012
6. Hitt, L.M., Wu, X.Z.D.J.: Investment in enterprise resource planning: business impact and productivity measures. J. Manag. Inf. Syst. **19**(1), 71–98 (2002)
7. Maguire, S., Ojiako, U., Said, A.: ERP implementation in Omantel: a case study. Ind. Manag. Data Syst. **110**(1), 78–92 (2010)
8. Liu, A.Z., Seddon, P.B.: Understanding how project critical success factors affect organizational benefits from enterprise systems. Bus. Process Manag. J. **15**(5), 716–743 (2009)

9. Dedrick, J., Gurbaxani, V., Kraemer, K.L.: Information technology and economic performance: a critical review of the empirical evidence. ACM Comput. Surv. CSUR **35**(1), 1–28 (2003)

10. Bharadwaj, A.S., Bharadwaj, S.G., Konsynski, B.R.: Information technology effects on firm performance as measured by Tobin's Q. Manag. Sci. **45**(7), 1008–1024 (1999)

11. Brynjolfsson, E., Hitt, L.: Paradox lost? firm-level evidence on the returns to information systems spending. Manag. Sci. **42**(4), 541–558 (1996)

12. Brynjolfsson, E., Hitt, L., Yang, S.: Intangible assets: computers and organizational capital Brook. Pap. Econ. Act. **2002**(1), 137–198 (2002)

13. Dewan, S., Min, C.: The substitution of information technology for other factors of production: a firm level analysis. Manag. Sci. **43**(12), 1660–1675 (1997)

14. Alter, S., Ginzberg, M.: Managing uncertainty in MIS implementation. Sloan Manage. Rev. **20**(1), 23–31 (1978)

15. Benaroch, M.: Managing information technology investment risk: a real options perspective. J. Manag. Inf. Syst. **19**(2), 43–84 (2002)

16. Boehm, B.: Software Risk Management. IEEE Computer Society Press, Los Alamitos (1989)

17. Keil, M., Cule, P.E., Lyytinen, K., Schmidt, R.C.: A framework for identifying software project risks. Commun. ACM **41**(11), 76–83 (1998)

18. Lyytinen, K., Mathiassen, L., Ropponen, J.: Attention shaping and software risk—a categorical analysis of four classical risk management approaches. Inf. Syst. Res. **9**(3), 233–255 (1998)

19. Dewan, S., Shi, C., Gurbaxani, V.: Investigating the risk-return relationship of information technology investment: firm-level empirical analysis. Manag. Sci. **53**(12), 1829–1842 (2007)

20. Reuters, T.: Skill shortage seen a risk for Indian IT sector growth. The Brunei Times, February 2015

21. Lichtenberg, F.R.: The output contributions of computer equipment and personnel: a firm-level analysis. Econ. Innov. New Technol. **3**(3–4), 201–218 (1995)

22. Anderson, M.C., Banker, R.D., Ravindran, S.: The new productivity paradox. Commun. ACM **46**(3), 91–94 (2003)

23. Zuboff, S.: In the age of the smart machine: the future of work and power. Basic Books, New York (1988)

24. Barua, A., Kriebel, C.H., Mukhopadhyay, T.: Information technologies and business value: an analytic and empirical investigation. Inf. Syst. Res. **6**(1), 3–23 (1995)

25. Mithas, S., Krishnan, M.S., Fornell, C.: Effect of information technology investments on customer satisfaction: theory and evidence. Ann Arbor, 1001 (2005)

26. Ravichandran, T., Lertwongsatien, C., Lertwongsatien, C.: Effect of information systems resources and capabilities on firm performance: a resource-based perspective. J. Manag. Inf. Syst. **21**(4), 237–276 (2005)

27. Kauffman, R.J., Walden, E.A.: Economics and electronic commerce: survey and directions for research. Int. J. Electron. Commer. **5**(4), 5–116 (2001)

28. Kulatilaka, N., Venkatraman, N.: Strategic options in the digital era. Bus. Strategy Rev. **12**(4), 7–15 (2001)

29. Sambamurthy, V., Bharadwaj, A., Grover, V.: Shaping agility through digital options: re conceptualizing the role of information technology in contemporary firms. MIS Q. **27**, 237–263 (2003)

30. Mithas, S., Tafti, A., Bardhan, I., Goh, J.M.: Resolving the profitability paradox of information technology: mechanisms and empirical evidence. University of Maryland Working Paper (2008)

31. Tanriverdi, H., Ruefli, T.W.: The role of information technology in risk/return relations of firms. J. Assoc. Inf. Syst. **5**(11–12), 421–447 (2004)
32. Fornell, C., Mithas, S., Morgeson III, F.V., Krishnan, M.S.: Customer satisfaction and stock prices: high returns, low risk. J. Mark. **70**(1), 3–14 (2006)
33. Kothari, S.P., Laguerre, T.E., Leone, A.J.: Capitalization versus expensing: evidence on the uncertainty of future earnings from capital expenditures versus R&D outlays. Rev. Account. Stud. **7**(4), 355–382 (2002)

Use of Ontologies in Information Systems Development

Osama Bassam J. Rabie[1,2] and Heinz Roland Weistroffer[1(✉)]

[1] School of Business, Virginia Commonwealth University, Richmond, VA 23284-4000, USA
{rabieob,hrweistr}@vcu.edu
[2] King Abdulaziz University, Jeddah 21589, Saudi Arabia

Abstract. Ontologies are becoming an essential component in many information technology (IT) applications, and systems development is an important and necessary step in the application of IT, which increasingly involves the Web. This article reviews the literature on the usefulness of ontologies in systems development with the purpose to stimulate discussion among information systems (IS) researchers and practitioners about the potential role of ontologies.

Keywords: Ontology · Systems development · Review

1   Introduction

There is an increasing awareness of the value of data, and computer supported techniques and tools are needed to process these data. Ontology is one type of tool that may support this process, and there is an increasing demand for research focused on ontology use [15]. This paper discusses the use of ontologies in systems development, with the goal to identify how ontologies can improve the development process in general, and Web development in particular.

Ontologies help convey information to systems developers and encourage them to follow specified standards. Ontologies may contribute to preventing the consequences of poor practices and thus negative impact on organizational image and budget.

Berners-Lee, who is credited with inventing the World Wide Web, stated "I have a dream for the Web [in which computers] become capable of analyzing all the data on the Web – the content, links, and transactions between people and computers. A 'Semantic Web', which should make this possible, has yet to emerge, but when it does, machines talking to machines will handle the day-to-day mechanisms of trade, bureaucracy and our daily lives. The 'intelligent agents' people have touted for ages will finally materialize" [1]. Ontologies provide a mechanism to build automated intelligence. They can facilitate handling big systems, automated tasks, knowledge processes, and decision-making, by formally representing domain knowledge in a way that software agents can understand [2]. Ontologies represent a set of concepts along with properties and relations within a specific domain, and thus can provide systems developers with the semantics of data that can be used with reasoning and problem-solving methods to develop systems with fewer other resources than would be required otherwise [3]. The diversity of ontologies allows systems from different domains to make use of ontology for systems development.

© Springer International Publishing Switzerland 2016
V. Sugumaran et al. (Eds.): WEB 2015, LNBIP 258, pp. 202–206, 2016.
DOI: 10.1007/978-3-319-45408-5_22

2 Use of Ontologies in Systems Development

Ontology usage in systems development may act as a guideline and shares several of the characteristics discussed by Oliveira et al. [4]. Ontology can describe appropriate actions and phases. The number of tools to handle ontology is increasing, and people with various levels of technical knowledge now commonly make use of ontologies. In the following sections we summarize the reviewed papers, classified by the specific purpose of the ontologies used in systems development.

2.1 Capturing Requirements

Farfeleder et al. [10] proposed a prototypic semantic guidance system to assist requirements engineers in capturing requirements. This system uses concepts, relations, and axioms of a domain ontology, which is used to assist in the requirements elicitation phase. It employs a semi-formal representation of the requirements, and the output is a list of suggested requirements that requirement engineers can use to build on and refine. The discussed use case consists of 43 requirements using natural language text. The types of requirements are functional requirements as well as requirements pertaining to safety, performance, reliability, availability, and cost. The results show that 36 out of the 120 attributes (30 %) were suggested by the semantic guidance system without manual change in the query or the ontology. 57.5 % of the attributes were partially suggested by the semantic guidance system. Overall, the semantic guidance system was able to help with 85 % of cases. The system was not able to suggest the number of attributes or understand words that were not included in the domain ontology. The ontology used can produce unrelated suggestions if it contains a too broad domain representation.

Kaiya and Saeki [11] propose an ontology based requirements elicitation (ORE) method, which uses domain ontology with inference rules and quality metrics to suggest requirements addition or deletion to enhance consistency. ORE works on functional requirements to assure its consistency, unambiguity, correctness, and completeness. In the reported study, three undergraduate students from a software engineering course were used as subjects, two being navigated subjects that used ORE for requirements elicitation, with the third student being free to use any method for requirements elicitation. The requirement elicitation was based on extending a given requirements list from a provided domain ontology, specifically music players. The ontology consisted of 48 concepts and 67 relationships written in Japanese, and the initial list of requirements included six items. The results show that the navigated subjects used simpler sentences than the other student, with fewer concepts mapped per requirement.

2.2 Detecting Conflicts in Change Requests

Liu and Yang [12] use an ontology-based blog to detect conflicts and inconsistencies in maintenance requirements. The proposal is to extend use-case and ontology. The use-case extension, shown in Fig. 1 is based on six nouns (role, function, data, activity, goal, and assumption) and five verbs (use, support, manipulate, achieve, and comply with).

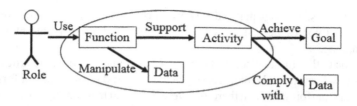

Fig. 1. The proposed use-case approach

A Taiwanese online newspaper with 1600 employees and a Taiwanese television station with 700 employees adopted the post-development change request management process and tried the blog-based prototype tool. The two companies posted a link to the proposed prototype tool on their web pages saying that they wanted to improve the websites. Several requirements were posted and many had conflicts.

2.3   Consistency Checking

Beydoun et al. [13] proposed three stages in a semantic-based approach to identify a suitable ontology for a system. First, the initial list of requirements are derived from documents and converted to ontology. Second, the ontology is used to check for requirements consistency. Finally, a set of reusable ontologies, relevant to system development, is identified and checked. The focus here is on the second stage, which uses rule inference and check requirements ontology for inconsistency, concept satisfaction, and subclasses relations. The method is implemented on a call management system, which handles the relationship between supervisors and employees (callers). Inconsistency is checked using rule inference. One issue found with the call management system was the assigning of a decency relation as an actor.

2.4   Requirements Management

Kumar and Kumar [14] introduce a framework for successful and efficient requirements management. The framework poses requirements management metrics to measure and manage software processes during the development of information systems. The framework was applied in the Indian Railway system and helped discover issues with communication and knowledge management, leading to improved project management activities.

3   Discussion

Our review shows that ontologies can represent knowledge in an adequate manner to eliminate or reduce ambiguity, vagueness, misinterpretation, and inconsistency. Ontologies can also help with information validation. In software development projects over distributed systems, ontology may reduce the number of requirements issued by management [14]. Ontologies help deal with uncertainty and incomplete information in decision-making.

In addition to the studies discussed in Sect. 2, we found several more articles dealing with ontologies in systems development, which however did not include information on application and performance and therefore were excluded from our review. Khemakhem et al. [5] proposed a Search Engine for Component based software development (SEC) that uses a discovery ontology and integration ontology. The discovery ontology provides system developers with a list of components to choose from, based on the developer's query and criteria. The integration ontology defines the internal structure of the components to assure that only integrated components are generated in the component list. Zhou et al. [6] proposed an Ontology-based PlaTform-specIfic software Migration Approach (OPTIMA). This framework develops a system application programming interfaces (APIs) ontology to assist in software migration. Lee and Zhao [7] proposed an ontology-based approach to elicit and analyze domain requirements. Domain ontology is created to assist in requirements elicitation using abstract stakeholders. Lee and Gandhi [8] proposed Ontology-based Active Requirements Engineering (Onto-ActRE). This framework actively adjusts the requirements based on stakeholders, domain problems, and goal-driven scenario compositions. And Dzung and Ohnishi [9] proposed an ontology-based reasoning method for requirements elicitation. Their method uses domain ontology with an inference engine to generate questions to ask stakeholders to cover each essential requirement. The inference also checks for requirements that are incorrect, redundant, and inconsistent.

4   Conclusion and Future Research

The main contribution of this review is that it provides an overview to future researchers of what has been done so far in investigating ontology use in systems development, and by extension, what gaps and opportunities exist for further research. The purpose is to stimulate discussion among information systems (IS) researchers and practitioners about the potential role of ontologies in the development of IS. As far as the authors are aware, no review on this topic has been published so far. While a few publications discuss ontology usage in general [15], this review explicitly focuses on ontology usage in systems development. The review found ontology to assist in improving systems quality, saving cost, enhancing system acceptance and understanding levels, and engaging users in organizational decision-making.

Future research may look at user involvement in ontology-based systems development, as there are no published studies dealing with this so far, though user involvement has been studied in other development approaches and would seem to be an important consideration (see for example [16]). In addition, future research may study the usage of ontology to address non-functional requirements. Most papers that propose ontology-based systems development address only functional requirements.

References

1. Berners-Lee, T.: Weaving the Web. Orion Business, London (1999)
2. Rabie, O., Norcio, A.F.: Discussion of some challenges concerning biomedical ontologies. In: Kurosu, M. (ed.) HCII/HCI 2013, Part II. LNCS, vol. 8005, pp. 173–180. Springer, Heidelberg (2013)
3. Schulz, S., Kumar, A., Bittner, T.: Biomedical ontologies: what part-of is and isn't. J. Biomed. Inform. **39**, 350–361 (2006)
4. Oliveira, T., Novais, P., Neves, J.: Development and implementation of clinical guidelines: an artificial intelligence perspective. Artif. Intell. Rev. **42**, 999–1027 (2014)
5. Khemakhem, S., Drira, K., Jmaiel, M.: SEC: a search engine for component based software development. In: Proceedings of the 2006 ACM Symposium on Applied Computing, pp. 1745–1750 (2006)
6. Zhou, H., Kang, J., Chen, F., Yang, H.: OPTIMA: an ontology-based PlaTform-specIfic software migration approach. In: Seventh International Conference on Quality Software (QSIC 2007), pp. 143–152 (2007)
7. Lee, Y., Zhao, W.: An ontology-based approach for domain requirements elicitation and analysis. In: First International Multi-Symposiums on Computer and Computational Sciences (IMSCCS 2006), pp. 364–371 (2006)
8. Lee, S.W., Gandhi, R.A.: Ontology-based active requirements engineering framework. In: 12th Asia-Pacific Software Engineering Conference (APSEC 2005) (2005)
9. Dzung, D.V., Ohnishi, A.: Ontology-based reasoning in requirements elicitation. In: Seventh IEEE International Conference on Software Engineering and Formal Methods, pp. 263–272 (2009)
10. Farfeleder, S., Moser, T., Krall, A., Ståalhane, T., Omoronyia, I., Zojer, H.: Ontology-driven guidance for requirements elicitation. In: Antoniou, G., Grobelnik, M., Simperl, E., Parsia, B., Plexousakis, D., De Leenheer, P., Pan, J. (eds.) ESWC 2011, Part II. LNCS, vol. 6644, pp. 212–226. Springer, Heidelberg (2011)
11. Kaiya, H., Saeki, M.: Using domain ontology as domain knowledge for requirements elicitation. In: 14th IEEE International Conference on Requirements Engineering, pp. 189–198 (2006)
12. Liu, C.-L., Yang, H.-L.: Applying ontology-based blog to detect information system post-development change requests conflicts. Inform. Syst. Front. **14**, 1019–1032 (2012)
13. Beydoun, G., Low, G., García-Sánchez, F., Valencia-García, R., Martínez-Béjar, R.: Identification of ontologies to support information systems development. Inform. Syst. **46**, 45–60 (2014)
14. Kumar, S.A., Kumar, T.A.: Study the impact of requirements management characteristics in global software development project: an ontology based approach. Int. J. Softw. Eng. Appl. **2**, 107 (2011)
15. Ashraf, J., Chang, E., Hussain, O.K., Hussain, F.K.: Ontology usage analysis in the ontology lifecycle: a state-of-the-art review. Knowl. Based Syst. **80**, 34–47 (2015)
16. Harris, M.A., Weistroffer, H.R.: A new look at the relationship between user involvement in systems development and system success. Commun. Assoc. Inform. Syst. **24**, 739–756 (2009)

Framework for Using New Age Technology to Increase Effectiveness of Project Communication for Outsourced IT Projects Executed from Offshore

Suparna Dhar[1](✉) and Indranil Bose[2]

[1] AIMA, Lodhi Road, New Delhi 110003, India
suparnadhar@gmail.com
[2] IIM Calcutta, Diamond Harbour Road, Kolkata 700104, India
bose@iimcal.ac.in

Abstract. Effective project communication involves interaction with diverse stakeholders having varied organizational and personal backgrounds. For offshore IT projects, cultural, language, time and geographic gap poses challenge in communication. Extending use of new technology in marketing communication, this paper proposes a conceptual framework to use new technology to increase effectiveness of project communication for remotely executed projects. It posits that social media powered by ubiquitous access using mobile and cloud computing, enriched by data analytics helps assuage communication challenges experienced by IT projects executed from offshore.

Keywords: Outsourced IT project · Offshore · Project communication · New age technology · Communication effectiveness

1 Introduction

Communication is the lifeblood of projects. Effective communication is essential to deal with problems related to uncertainty and interdependence in IT projects involving diverse stakeholders (McKay et al. 2014). Adequate and appropriate communication with stakeholders is vital in conceptualization, specification, design and development of IT artifacts (McKay et al. 2010). But effective communication continues to be one of Project Managers primary challenges (Muszynska 2015). This challenge is exacerbated by time and cultural gap between stakeholders, in case of remotely executed IT offshore projects. Conventional communication channels fail to create personal and emotional connect in virtual project teams and trust between vendor and client. Mobile and social media gained high adoption globally. Cloud computing, data analytics, augmented reality (AR) and ubiquitous access have found application in content enrichment. Marketers are combining the inherent power of these technologies for effective product promotion (Chau and Xu 2012; O'Mahony 2015).

Research Question: Can application of new age technology help remotely executed IT offshore projects overcome the communication challenges?

© Springer International Publishing Switzerland 2016
V. Sugumaran et al. (Eds.): WEB 2015, LNBIP 258, pp. 207–211, 2016.
DOI: 10.1007/978-3-319-45408-5_23

This paper explores the options to increase effectiveness of project communication using new age technology taking cues from marketing.

2   Literature Review

Effective Communication: The topic of effective communication is of interest since Aristotle proposed his communication model. Transmission model by Shanon and Weaver (1949), Communication reaction model by Schramm (1954), Source message channel receiver (SMCR) model by Berlo (1960), Riley's double vortex model by Riley and Chernatony (1967), Cambell and Level (1985) and Layer based pragmatic communication model by Targowski and Bowmaan (1988) are some of the prominent communication models prevalent today (Lee 1993). These models study exchange of information in a communication context, the attributes that influence communication and challenges in the process. McKay et al. (2010) argued that meanings, information and communication are all context-dependent and effectiveness of communication depends how artifacts correspond in particular contexts. With advances in communication technology, organizations have wider choice of communication platforms and channels for faster and richer information exchange. In the business context, Galle's model views development of an artifact (McKay et al. 2010) as a result of communication of ideas, specifications, representations prior to development of the final artefact, and stakeholders' interpretation of information.

Challenges in Outsourced IT Projects Executed from Offshore: Nearly half of offshore outsourced initiatives "fail" or do not meet stated performance objectives (Nakatsu and Iacovou 2009). Inability to manage relationship, information constraints, geographic distance, cultural differences, language barrier, time zone difference, inefficient multi-party coordination, inadequate user involvement contribute to risk of failure of offshore outsourced projects (Langer et al. 2014). Client-provider interactions can go beyond contract, rules and agreements, and rest on trust and commitment (Lee et al. 2003). Understanding of each other's business to set realistic expectations, sharing risks, benefits and mutually beneficial standards help nurture vendor-client relationship. In IT projects, communication extends to customers' customer as well as other projects (McKay et al. 2014). Virtual projects teams, located across geographies, organizations and time zones, service these project. With time, service landscape of offshore IT projects is changing from regular services to niche services involving transformation, innovation and developing customized solutions. The shift to niche services endows higher complexity and lower familiarity. Best practices, processes and frameworks offered by traditional project management theory cannot be tailored to resolve the unexpected and difficult situations arising in such scenarios.

Use of New Age Technology in Marketing: Mobile phones are equipped with camera, scanners and GPS with push and pull capability, making it a rich medium for customer engagement (Yadav et al. 2015). Constant companionship aspect of mobile facilitates immediacy and quality of communication. To make purchase decisions, consumers look for information about the products or services on Social Networking Sites where they get trusted feedback from other consumers (Bashar et al. 2012). Content sharing sites like "YouTube" serve as an alternate channel for advertisement. Blogs and micro-blogging sites like "Twitter" empower consumers to share information, which makes it a gold mine for business intelligence (Chau and Xu 2012). Social media on mobile allows real time user generated content that enables location and time based customer engagement. Cloud enables flexible, market-oriented services whose success depends on quick response to rapidly changing customer needs, through collaboration between and among organizations (Mansuri et al. 2014). Search analytics, consumer behavior analytics, customer preferences, blog content analysis and location based analytics help targeted product promotion. Ubiquitous access and immediacy helps overcome friction in the commerce chain across borders, cultural differences, geography and cognitive impediments (McGuigan and Manzerolle 2014). AR integrates virtual information into a user's physical environment (O'Mahony 2015). With AR, mobile phone apps make a product 'come alive', to change colors of a shoe or take a virtual tour of a house.

3   Methodology

Literature review forms the basis of the proposed conceptual framework extending the application of new age communication technology in marketing communications to offshore project communication to increase its effectiveness. Since Berlo's SMCR model is most frequently referred by scholars (Turitsa 2013), this research uses SMCR model. Case studies and contemporary research findings are used to validate the efficacy of it. Google Scholar search was used to locate relevant literature.

4   Proposed Conceptual Framework

This section proposes the conceptual framework (Fig. 1) to increase effectiveness of project communication and assuage communication barriers in outsourced IT projects executed from offshore by using new age technology. Platforms like social media provide an opportunity for self-revelation that helps build trust between people located

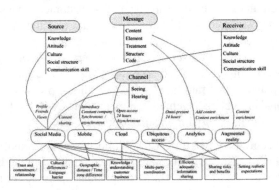

Fig. 1. Proposed conceptual framework using SMCR model to map communication challenges in offshore IT projects (rectangle) to technology (rounded rectangle).

remotely. Sharing enriched content helps overcome language barriers and improves knowledge sharing. Asynchronous and omni-present communication channels reduce time barriers in communication and help better risk containment.

Validation of the Proposed Conceptual Framework: In disaster relief project in response to 2010 Haiti earthquake new age technology was used extensively to orchestrate relief operations across functions and organization boundaries (Yates and Paquette 2011). Like offshore IT projects, the situation was complex, and involved multiple stakeholders across multiple organizations, having diverse cultural background. Social media platform was used for open online exchange of information that helped eliminate conflicting information flows, helping faster decision making necessary to prevent losses. It served as a platform for generation, dissemination, sharing and editing/refining of content, conversation and interaction spanning organizational and geographical boundaries. Wikis were used to share new updates. Cloud computing environment enabled collaborative management of documents, spreadsheets, photos and videos, bringing together ad hoc networks with varied expertise and context. Ubiquitous access was used to make communications from anywhere with internet access. Data visualization techniques using Google maps, helped create a common context between decision makers. Identity Management case study (DiMicco and Millen 2007) finds Facebook being increasingly used in workplace to keep in touch and increase awareness and contact with distant employees. The act of "people sensemaking" helps improve interpersonal communication. Case study on micro-blogging in the enterprise (Zhang et al. 2010) indicates that it addresses diverse communication needs across organization culture in work environment. Agile distributed software development case study (Estler et al. 2012) proposes use of cloud in software development for real-time concurrent work. Supported by case studies, Wang et al. (2012) posits the efficacy of AR in collaboration in artifact development and e-learning in virtual environments.

SNS provides an open platform for large scale knowledge sharing, distributing, networking and socialization of knowledge (Ellison et al. 2014). It serves as informal channel to share emotions among distributed team members (Muszynska 2015). Projects have an emotional process, involving personal commitment of team members and empathy, trust and confidence within team and beyond (McKay et al. 2014). Social media serves as the virtual water cooler, to support informal and serendipitous interactions that help co-creation and value generation across project, users and stakeholders separated by time and culture.

5   Conclusion

The proposed conceptual framework ties contemporary communication barriers in outsourced IT project executed from offshore to attributes of new age communication technology. It will help practitioners formulate project communication strategy using new communication channels. Future research should attempt to test the framework using experimental research design. The framework can be extended to remotely executed projects across sectors and geographies and other communication models.

References

Bashar, A., Ahmad, I., Wasiq, M.: Effectiveness of social media as a marketing tool: an empirical study. Int. J. Mark. Fin. Serv. Manage. Res. 1(11) (2012)

Berlo, D.: The Process of Communication. Holt, Rinehart and Winston Inc., New York (1960)

Chau, M., Xu, J.: Business intelligence in blogs: understanding consumer interactions and communities. MIS Q. **36**(4), 1189–1216 (2012)

DiMicco, J.M., Millen, D.R.: Identity management: multiple presentations of self in facebook. In: Proceedings ACM Group Conference, Florida, USA (2007)

Ellison, N., Gibbs, J.L., Weber, M.S.: The use of enterprise social network sites for knowledge sharing in distributed organizations the role of organizational affordances. Am. Behav. Sci. **59**(1), 103–123 (2014)

Estler, H.-C., Nordio, M., Furia, M., Meyer, B., Schneider, J.: Agile vs. structured distributed software development: a case study. In: Proceedings of the 7th International Conference on Global Software Engineering, IEEE 2012, pp. 11–20 (2012)

Langer, N., Slaughter, S.A., Mukhopadhyay, T.: Project managers' practical intelligence and project performance in software offshore outsourcing: a field study. Inform. Syst. Res. **25**(2), 364–384 (2014)

Lee, D.: Developing effective communications. university of missouri extension (1993). http://extension.missouri.edu/publications/DisplayPub.aspx?P=CM109. Accessed 06 June 2015

Lee, J., Huynh, M.Q., Kwok, R.C., Pi, S.: IT outsourcing evolution—: past, present, and future. Commun. ACM **46**(5), 84–89 (2003)

Mansuri, A., Verma, M., Laxkar, P.: Benefit of cloud computing for educational institutions and online marketing. Inform. Secur. Comput. Fraud. **2**(1), 5–9 (2014)

McGuigan, L., Manzerolle, V.: All the world's a shopping cart: theorizing the political economy of ubiquitous media and markets. New Media Soc. **17**, 1830–1848 (2014)

McKay, J., Grainger, N., Marshall, P., Hirschheim, R.: Artefaction as communication: redesigning communication models. In: 2010 Proceedings of the Paper Presented at the ACIS, Brisbane, Australia, January 2010

McKay, J., Marshall, P., Grainger, N.: Rethinking communication in IT project management. In: 2014 47th Hawaii International Conference on System Science Paper Presented at the (HICCS), Hawaii, USA, January 2014

Muszynska, K: Communication management in project teams – practices and patterns. In: Joint International Conference on Paper Presented at MakeLearn, Bari, Italy, May, 2015

Nakatsu, R.T., Iacovou, C.L.: A comparative study of important risk factors involved in offshore and domestic outsourcing of software development projects: a two-panel Delphi study. Inform. Manage. **46**, 57–68 (2009)

O'Mahony, S.: A proposed model for the approach to augmented reality deployment in marketing communications. Procedia Soc. Behav. Sci. **175**, 227–235 (2015)

Targowski, A.S., Bowman, J.P.: The layer-based, pragmatic model of the communication process. Int. J. Bus. Commun. **25**, 15–24 (1988)

Wang, X., Love, P., Klinc, R., Kim, M., Davis, P.: Integration of E – learning 2.0 with Web 2.0. J. Inform. Technol. Constr. **17**, 387–396 (2012)

Yadav, M., Joshi, Y., Rahman, Z.: Mobile social media: the new hybrid element of digital marketing communications. Procedia Soc. Behav. Sci. **189**, 335–343 (2015)

Yates, D., Paquette, S.: Emergency knowledge management and social media technologies: a case study of the 2010 Haitian earthquake. Int. J. Inform. Manage. **31**, 6–135 (2011)

Zhang, J., Qu, Y., Cody, J., and Wu, Y.: A case study of microblogging in the enterprise: use, value, and related issues. In: Proceedings of CHI, pp. 243–252. ACM Press (2010)

Does Too Much Regulation Kill the Online Gambling Industry?: An Empirical Analysis of Regulation Effects Using VAR Model

Moonkyoung Jang[1], Seongmin Jeon[2(✉)], Byungjoon Yoo[1],
Jongil Kim[3], and Changhee Han[4]

[1] Graduate School of Business, Seoul National University, Seoul, Korea
{moonk14,byoo}@snu.ac.kr
[2] Business School, Gachon University, Seongnam, Gyeonggi-do, Korea
smjeon@gachon.ac.kr
[3] School of Law, Seoul National University, Seoul, Korea
thisishim@snu.ac.kr
[4] School of Business Administration, Hanyang University, Seoul, Korea
chan@hanyang.ac.kr

Abstract. Motivated by the growing interest of online gambling regulation, we empirically investigate whether the regulation successfully decreases online gambling addiction. For this purpose, we use a vector autoregression (VAR) model to identify interrelationship changes among online gambling games and service platforms due to the regulation. In particular, we propose three theoretical perspectives: the role of prior experience, switching costs and network externalities. We find that the impact size of the regulation is different depending on levels of prior experience, and the regulation significantly affects the switching behavior of users on online gambling platform. Therefore, we offer one of the first empirical evidences that examine the regulation effects on online gambling using VAR model. We also suggest the policy makers should make suitable regulations for each user group to effectively avoid generating gambling addicts without interrupting the economic growth of the online gambling industry.

Keywords: Online gambling · VAR · Vector autoregression · Regulation effect

1 Introduction

Online gambling is an emerging industry with great prospect but many governments are trying to control online gambling despite its industrial importance [1]. At the same time, they have started to pay attention to the potential of legalized online gambling as a good source of tax revenue as well. Thus, they try to ensure that online gambling is fairly taxed without any interruption in its economic growth. Therefore, we apprehend that the original intent of the regulation is to prevent an increase in the incidence of gambling addiction while the market size of the online gambling industry is still increasing. However, we doubt whether the governments achieve their original purpose or not. The reason is the governments normally legislate to control online gambling gaming without

© Springer International Publishing Switzerland 2016
V. Sugumaran et al. (Eds.): WEB 2015, LNBIP 258, pp. 212–216, 2016.
DOI: 10.1007/978-3-319-45408-5_24

considering the various characteristic of users on each online gambling gaming. To achieve the original intent of the regulation, the regulation should control the gambling behaviors of potential gambling addicts only, while not restricting those of casual users who use online gambling as a form of entertainment. In this context, many researchers have published valuable papers offering recommendations for effective regulations for online gambling. However, quantitative evidence is limited because most of attempts are made through a case study approach. In response to these limitations of previous studies, we empirically investigate whether the regulation achieves its original purpose using the VAR model with longitudinal data on online gambling games. We expect to find the hidden impacts of the regulation using three theoretical perspectives: the role of prior experience, switching costs, and network externalities.

2 Background

2.1 The Role of Prior Experience

The technology acceptance model (TAM) is one of the most significant research models on information system (IS) acceptance and continuance [2]. It suggests two main variables: perceived usefulness and perceived ease of use. To understanding IS acceptance behavior in non-work environments, Van der Heijden [3] expended the TAM model by adding a new construct, perceived enjoyment. Therefore, the extended TAM model can be applied to this context since online gambling games are mainly played for hedonistic or entertainment purposes in an online platform. The moderating effect of prior experience on IS acceptance or continuance has also been studied by many IS researchers. For example, Kim et al. [4] find that perceived enjoyment has a great impact on the use intention of experienced users than on that of inexperienced users. In the case of online gambling, software-generated gambling requires much time for players to develop gambling skills [5], and users normally play high-level games after playing skillfully on low-level games. Thus, we assume that users of high-level games have more online gambling experience than do users of low-level games. In addition, the regulation may affect users' attitude toward online gambling because it directly decreases website accessibility and the perceived enjoyment of users. We therefore hypothesize:

Hypothesis 1. The regulation related to online gambling will have larger effect on the usage of high-level games than on that of low-level games.

2.2 Switching Costs

Users of online gambling normally tend to play on only one platform, because their in-game money or level cannot be transferred to other platforms. This kind of user behavior can be explained by the concept of switching cost, which indicates the form of transaction costs including learning costs, switching brands, or contractual costs [6]. The regulation may decrease users' perceived switching cost because it encourages users to change platforms by setting game money restrictions on only one platform. In this situation, they can simply avoid the game money restriction by changing their service platform when they reach the restriction on one platform. Based on this, we hypothesize:

Hypothesis 2. The regulation on online gambling will be negatively related to perceived switching cost of users.

2.3 Network Externalities

An online gambling system basically provides a network that connects users to play against other users [7]. The value of the online gambling system increases for users as the number of other users in the same network increase. This can be described by the concept of network externalities, which indicates the value of a certain technology increases with the number of other users who are connected to the same system [8]. However, the regulation seems not to consider the characteristic of network externalities in online gambling. The government typically creates legislating to control online gambling platforms without considering the number of users in each system. We suppose that legislators overlook network externalities as an essential factor to determine the impact of the regulation. Based on this, we hypothesize:

Hypothesis 3. The impact size of the regulation related to online gambling will be different depending on the number of users in online gambling system.

3   Research Method

To analyze the regulation effect on online gambling gaming, we acquire panel data on online poker gaming from Gametrics, one of the major online gaming data providers in Korea. The Korean government considers online gambling to be the main cause of generating gambling addiction, so it publicized a detailed regulation proposal on November 29, 2012. This new regulation would have a negative impact on the Korean online gambling industry, because many Korean users may move to foreign online gambling platforms or reduce their overall gaming time due to this new regulation. The dataset consists of a balanced panel dataset of game playtime for the period of December 2010 to November 2014. The original dataset includes daily playtime per game and we sum this into weekly data because people tend to play online poker games on a weekly basis as with other entertainment [9]. In terms of game usage, the market share of the top five games is about 75 %. The remainder is shared among seven companies, each of which has less than a 5 % share. Since the influence of these companies on the Korean online poker industry is too small to investigate, we decide to use the dataset of the top five poker games: *Hangame lowbaduki-poker, Netmarble poker, Hangame highlow-poker, Hangame seven-poker, Pimang lowbaduki-poker.* We investigate the dataset using VAR model to analyze the influence of regulation on each online gambling gaming. Through this model, we can capture the linear interdependencies among multiple time series and compare the size variance of a ripple effect on playtime of each game due to the regulation.

4 Empirical Result

We examine the results of Granger causality and impulse response function (IRF) which are reported as a standard practices for VAR model [10].

To test hypothesis 1, we divide our dataset into two groups based on game level. We assume that the users of high-level gambling have more gambling experience than the users of low-level gambling, since software-generated gambling requires time to develop gambling skills. Thus, we categorize *Hangame lowbaduki-poker*, *Hangame highlow-poker* and *Pimang lowbaduki-poker* as high-level games and *Hangame seven-poker* and *Netmarble poker* as low-level games. We then conduct dynamic forecasting to comparing forecasted and actual playtime of each level. For the high-level games, the forecasted playtime is less than the actual playtime. In contrast, for the low-level games, the forecasted playtime is greater than the actual playtime. According to the Granger causality test, the game usage of low-level games is the Granger cause of the game usage of high-level games before the regulation. This is quite normal, because users play low-level games first and then move on to high-level games. However, after the regulation, the usage of low-level games is the Granger cause of the usage of high-level games and vice versa. This does not seem normal, because most users increasingly seek to enjoy high-level games as they get used to the online gaming experience, rather than returning to low-level games. IRFs show that the usage change of low-level games has a greater impact on high-level games. Furthermore, the impact maintains longer after the regulation.

To test hypothesis 2 and hypothesis 3, we classify five poker games by service platforms to investigate the effect of the regulation on users' switching behavior on gambling websites. We first conduct dynamic forecasting to identify the overall trend of total playing time and to compare forecasted and actual playtime. According to the dynamic forecasting, playtimes on all platforms are forecasted as greater than actual playtime, except in the case of *NHN*, which has the largest market share in online gambling industry. The playtime of *NHN* is forecasted at a negative value after February 2014 but it still maintains the number one position. This appears to be evidence in support of our second hypothesis, because *NHN* users switch less than the users of other platforms after the regulation. Therefore, we can conclude that the larger service platform has a lower impact on the regulation. The Granger causality test shows that *CJ* becomes the Granger cause of the other companies after the regulation. The result of the IRF test illustrates that the impact size of *CJ* is bigger and that the impact sustains longer after the regulation.

5 Conclusion

This paper empirically investigates the effect of the regulation on online gambling from three theoretical perspectives: the role of prior experience, switching cost, and network externalities. We find that the regulation decreases total playtime and significantly changes the correlation of playtime of each online gambling games. However, we cannot simply conclude that decreased playtime means the regulation is successful. This is because the object of the regulation is only to reduce the occurrence of gambling addiction, not to cause the online gambling industry to decline. It is not sufficient to see decreased playtime in order to fully understand the hidden impact of the regulation. Thus, we design the VAR model as a matching methodology given the theory, assumption and available data. Specifically, we use both the Granger causality test and IRF test to confirm the relationship among variables. Our test results show that the regulation affects online gambling in many ways. Therefore, we discover that the regulation has a more sustained effect on players with more gaming experience. In addition, it significantly increases the players' chance of changing service platforms. We believe this research identifies an interesting and significant phenomenon that has not yet received much study. For future research, the current results can be enriched by incorporating individual level data of each game. We expect that this study has potential implications for IS researchers and policy makers.

References

1. Stewart, D.O., Gray, L.: Online gambling five years after UIGEA. American Gaming Association (2011)
2. Davis, F.D.: Perceived usefulness, perceived ease of use, and user acceptance of information technology. MIS Q. **13**, 319–340 (1989)
3. Van der Heijden, H.: User acceptance of hedonic information systems. MIS Q. **28**, 695–704 (2004)
4. Kim, B., Choi, M., Han, I.: User behaviors toward mobile data services: the role of perceived fee and prior experience. Expert Syst. Appl. **36**, 8528–8536 (2009)
5. Griffiths, M.: Internet gambling: issues, concerns, and recommendations. CyberPsychol. Behav. **6**(6), 557–568 (2003)
6. Klemperer, P.: Markets with consumer switching costs. Q. J. Econ. **102**, 375–394 (1987)
7. Sieroty, C.: Panel says critical mass matters for online poker. Las Vegas Business Press (2011)
8. Katz, M.L., Shapiro, C.: Network externalities, competition, and compatibility. Am. Econ. Rev. **75**, 424–440 (1985)
9. Ma, X., Kim, S.H., Kim, S.S.: Online gambling behavior: the impacts of cumulative outcomes, recent outcomes, and prior use. Inf. Syst. Res. **25**, 511–527 (2014)
10. Stock, J.H., Watson, M.W.: Vector autoregressions. J. Econ. Perspect. **15**, 101–115 (2001)

Developing and Evaluating a Readability Measure for Microblogging Communication

Marten Risius(✉) and Theresia Pape

Institute for Information Systems, Goethe University Frankfurt,
Theodor-W.-Adorno-Platz 4, 60329 Frankfurt am Main, Germany
risius@wiwi.uni-frankfurt.de, theresia.pape@gmail.com

Abstract. Especially due to the recent expansion of social media platforms, researchers and practitioners exert ever growing efforts to advance big data analytics techniques to derive actionable insights from social networks. Although a substantial body of research has shown that readability is of vital importance for the success of text-based communication, currently it is rarely considered in social media research or especially microblogging. In this research project, we intend to develop a readability measure for microblogging messages that is applicable to large scale data analysis and also provides concrete formulation recommendations for single messages. We will combine text mining and machine learning techniques to analyze a sample of approximately 6.8 million Twitter messages from and about 33 large S&P 100 companies to develop a respective readability measure.

Keywords: Readability · Microblogging · Big data analytics · Social media analytics

1 Introduction

With the proliferation of social media platforms, companies greatly rely on these new channels for various business purposes like branding activities, customer service & support or talent acquisition [1]. In this regard, Twitter has been shown to be a viable solution for E-Commerce activities by enabling organizations to successfully communicate with customers [2] and draw conclusions about users' opinions [3].

Although the relevance of readability has long been established in various business areas [4, 5], it is neglected in contemporary text mining applications especially in microblogging [6, 7]. Readability is understood as the effort necessary to comprehend a text. It generally depends on textual features (e.g., word frequency, sentence length, and lexical density) and reader characteristics (e.g., age, level of education) [8].

With our research project, we address this big data analytics research gap by deploying machine learning and text mining techniques to develop a readability measure that is applicable to the largescale analysis of microblogging posts and simultaneously provides formulation recommendations to properly adjust message readability. Thereby, we address research questions of what constitutes readability in microblogging communication and how does it translate into business value?

© Springer International Publishing Switzerland 2016
V. Sugumaran et al. (Eds.): WEB 2015, LNBIP 258, pp. 217–221, 2016.
DOI: 10.1007/978-3-319-45408-5_25

2   Theoretical Background

In this work we understand readability in terms of the reading ease that supports the comprehension and memorization of textual material. Readability is conceptualized as the effort necessary for a person of a particular age and educational level to understand a text and is commonly operationalized in school grade levels [4, 9]. Measures of readability have predominantly been applied by pedagogues to appropriately design text-books or by journalists to align their style of writing [7]. For example, the London Times address an educational level of a 12th grade student while the Washington Post and the New York Times target as 10th grade level [8]. Moreover, readability has been found to exert business value in the sense that a higher readability of a companies' annual reports translates into larger financial investments into the companies over a period of 13 years [5].

2.1   Relevance of Readability in E-Commerce

While research around readability has a long history and brought forth a substantial body of insights and measures, readability tests are of unprecedented popularity [8]. Recently the concept of readability has been introduced to the field of E-Commerce where it has been found to be a major contributing factor for content quality on E-Commerce websites in terms of characteristics and presentation of information [10, 11]. With the ever-growing proactive online search behavior of users for opinions, information quality is commonly considered to be of peculiar relevance in E-Commerce [12]. Especially in the case of social media enabled E-Commerce, consumers proactively seek opinions online to make judgements ranging from low-involvement (e.g., choice of films) to high-involvement decisions (e.g., stock investments) [4]. The quality of content affects organizational success through the satisfaction of website users [11], website usability and consumer trust [13], as well as the customer attitude towards websites [14]. Accordingly, in the specific case of E-Commerce related social media content, initial research has found readability to be a major predictor of online review manipulation [4].

2.2   Readability in Social Media Communication and Microblogging

In this study, we generally assume that the user's ability to understand a message is a decisive factor for the message's relevance [15]. Accordingly, researchers were able to identify high quality answers based on readability measures [16] and improve search engine functionality by customizing the readability of its search results [15]. However, it needs to be acknowledged that microblogging communication – especially on Twitter – has been found to linguistically differ (i.e., more conservative and formal) from other means of short text communication (e.g., online chat and SMS) [6, 17]. Currently only preliminary research has begun to context-specifically apply the concept of readability to Twitter communication. The currently single large scale analysis of tweet readability applied a basic adaptation of the common Flesch-Reading-Ease formula [18] and provided initial evidence for the validity of tweet readability in terms of correlations with the grad-uation rate in the U.S. [6]. However, the authors acknowledge certain validity limitations of the common readability measure which does not consider context-specific linguistic styles. Thus, we consider it necessary to develop a readability measure that is appropriate for the specific microblogging environment or singular E-Commerce purposes but is simultaneously also suitable for big data analysis.

3   Empirical Study

In order to develop a readability measure that is appropriate for different E-Commerce purposes on Twitter but also applicable to largescale message analysis, we conducted a preliminary text-mining analysis and intend to expand our respective research in future. Out of the different E-Commerce platforms worth analyzing regarding our research questions, we chose Twitter which offers a public data access [19] and is accepted by companies and customers for marketing activities [2, 3].

3.1   Future Research Approach

Since the readability indices applied in existent microblogging studies did not consider specificities of the tweets themselves and were intended for other types of texts [7], we reviewed readability measures to find a more appropriate conceptualization which also considers singular words. In this regard, we see the "New Dale-Chall Readability Formula" as the most appropriate starting point for a conceptual basis [20]. This formula applies among others a list of 3,000 familiar words to estimate readability. It is based on the assumption that it is more the use of familiar words – rather than the simple number of syllables or letters – which determines the ease of comprehension. As such it exceeds other common readability formulas by not only considering the average sentence length but also the percentage of difficult words within a given text. The new Dale-Chall formula was well-validated against a wide variety of criteria [8].

Nonetheless, in our further research process we intend to move away from the classic few feature readability measures towards the modern machine learning based readability approaches capable of considering large numbers of relevant text features [7]. Generally, we plan to follow four different steps to expand from the classic readability formula and ultimately validate our measure. First, we will conduct a latent semantic text mining analysis separately for the different company account categories (e.g., branding, sales, and customer service and support) and industry sectors (e.g., finance sector, consumer goods, and manufacturing) in our data sample to identify the particular set of familiar and non-familiar words. Second, readability measures are commonly standardized by levels of age and education. Thus, we apply a crowdsourcing approach to obtain respective readability ratings of tweets and establish a gold standard criterion necessary for our subsequent analytical step. The coding will be conducted according to the five-step manual content analysis procedure [21]. Third, in combination with the previously developed familiar word lists and the manually coded readability measure, a machine learning algorithm will be deployed to identify the influence of different factors that determine readability. In the last step, we analyze how the different levels of readability translate into respective success measures of word of mouth and attitudinal loyalty [22].

Through this comprehensive multi-step process that combines qualitative and text-mining techniques, we address common concerns of readability formulas regarding reliability and comparability [23]. This approach enables us to develop readability measures by relying on large and context-specific datasets that allow timely word list updates as well.

4   Conclusion and Future Research

The goal of this project is to develop a readability measure that is applicable to micro-blogging posts and can support companies' E-Commerce activities on Twitter. This readability measure is intended to serve two different purposes. On the one hand, it provides insights through big data social media analytics and, on the other hand, it makes formulation recommendations for single messages.

This readability approach will be able to characterize users and identify target-group members. Considering that readability measures are standardized relative to age and educational level, analyzing user messages of the companies' audience provides respective insights into the characteristics of its followers. Thereby, companies are enabled to further customize their communication and marketing strategies.

By deriving lists of (non-)familiar words in certain semantic contexts, we will be able to provide specific synonymous alternatives that in- or decrease the readability of messages relative to the target group. These lists can be adapted specifically to certain industry branches (e.g., sports, finance, consumer goods, industry) and account types (e.g., branding, sales, customer service and support).

References

1. Risius, M., Akolk, F.: Differentiated sentiment analysis of corporate social media accounts. In: Proceedings of the 21st American Conference on Information Systems (AMCIS) (2015)
2. Burton, S., Soboleva, A.: Interactive or reactive? Marketing with twitter. J. Consum. Mark. **28**, 491–499 (2011)
3. Bulearca, M., Bulearca, S.: Twitter: a viable marketing tool for SMEs. Glob. Bus. Manag. Res. Int. J. **2**, 296–309 (2010)
4. Hu, N., Bose, I., Koh, N.S., Liu, L.: Manipulation of online reviews: an analysis of ratings, readability, and sentiments. Decis. Support Syst. **52**, 674–684 (2012)
5. Loughran, T., Mcdonald, B.: Measuring readability in financial disclosures. J. Finance **69**, 1643–1671 (2014)
6. Davenport, J.R., Deline, R.: The Readability of Tweets and Their Geographic Correlation with Education. arXiv preprint arXiv:1401.60582014
7. Temnikova, I., Vieweg, S., Castillo, C.: The case for readability of crisis communications in social media. In: Proceedings of the 24th International Conference on World Wide Web Companion, pp. 1245–1250. International World Wide Web Conferences Steering Committee (2015)
8. Dubay, W.H.: The Principles of Readability. Online Submission (2004)
9. Zakaluk, B.L., Samuels, S.J.: Readability: Its Past, Present, and Future. ERIC (1988)
10. Rababah, O.M.A., Masoud, F.A.: Key factors for developing a successful E-commerce website. Commun. IBIMA **2010**, 1–9 (2010)

11. Zhang, X., Keeling, K.B., Pavur, R.J.: Information quality of commericial web site home pages: an explorative analysis. In: 21st International Conference on Information Systems (ICIS), pp. 164–175. Association for Information Systems (2000)
12. Molla, A., Licker, P.S.: E-commerce systems success: an attempt to extend and respecify the delone and maclean model of is success. J. Electron. Commer. Res. **2**, 131–141 (2001)
13. Pelet, J.-E., Papadopoulou, P.: The effects of e-commerce websites colors upon consumer trust. In: 38th European Marketing Academy Conference (EMAC) (2009)
14. Chen, Q., Wells, W.D.: Attitude toward the site. J. Advertising Res. **39**, 27–37 (1999)
15. Kim, J.Y., Collins-Thompson, K., Bennett, P.N., Dumais, S.T.: Characterizing web content, user interests, and search behavior by reading level and topic. In: 5th ACM International Conference on Web Search and Data Mining (WSDM), pp. 213–222. ACM (2012)
16. Agichtein, E., Castillo, C., Donato, D., Gionis, A., Mishne, G.: Finding high-quality content in social media. In: 2008 International Conference on Web Search and Data Mining, pp. 183–194. ACM (2008)
17. Hu, Y., Talamadupula, K., and Kambhampati, S.: Dude, srsly?: The surprisingly formal nature of twitter's language. In: 7th International Conference on Weblogs and Social Media (ICWSM). AAAI (2013)
18. Flesch, R.: A new readability yardstick. J. Appl. Psychol. **32**, 221 (1948)
19. Bruns, A., Stieglitz, S.: Towards more systematic twitter analysis: metrics for tweeting activities. Int. J. Soc. Res. Methodol. **16**, 91–108 (2013)
20. Chall, J.S., Dale, E.: Readability Revisited: The New Dale-Chall Readability Formula. Brookline Books, Cambridge (1995)
21. Morris, R.: Computerized content analysis in management research: a demonstration of advantages & limitations. J. Manag. **20**, 903–931 (1994)
22. Risius, M., Beck, R.: Effectiveness of corporate social media activities in increasing relational outcomes. Inf. Manag. **52**, 824–839 (2015)
23. Stevens, K.C.: Readability formulae and mccall-crabbs standard test lessons in reading. Reading Teach. **33**, 413–415 (1980)

Sharing Behavior in Online Social Media: An Empirical Analysis with Deep Learning

Donghyuk Shin[1](✉), Shu He[1], Gene Moo Lee[2], and Andrew B. Whinston[1]

[1] University of Texas at Austin, Austin, TX 78712, USA
dshin@cs.utexas.edu, shuhe@utexas.edu, abw@uts.cc.utexas.edu
[2] University of Texas at Arlington, Arlington, TX 76019, USA
gene.lee@uta.edu

Abstract. We conduct a large-scale empirical study on the sharing behavior in social media to measure the effect of message features and initial messengers on information diffusion. Our analysis focuses on messages created by companies and utilizes both textual and visual semantic content by employing state-of-the-art machine learning methods: topic modeling and deep learning. We find that messages with multiple conspicuous images and messengers with similar content are crucial in the diffusion process. Our approach for semantic content analysis, particularly for visual content, bridges advanced machine learning techniques for effective marketing and social media strategies.

Keywords: Social media · Information diffusion · Deep learning · Topic modeling · Community analysis

1 Introduction

Online social networks have emerged as one of the most important media platforms for companies to deliver their messages to customers. Effectively delivering messages on social media is one of the major marketing issues. There are mainly two crucial aspects to consider in effective communication: (1) articulating the *messages* in the right format; (2) selecting the right initial information injection points, i.e., *messengers*, that would maximize the information diffusion process.

Previous research on messages mostly focused on text analysis [5,11]. While there is a rapid increase of visual contents in social media [6], only a few studies incorporate visual aspects of the message, where only simple characteristics are utilized [7,8]. In this paper, we take a major step forward and adopt state-of-the-art deep learning approaches to extract meaningful features from visual contents in social media. Existing literature has revealed that the messengers' characteristics such as position and tie strength have significant influence on the diffusion process [1,10]. We follow this direction to incorporate the network position of the initial messenger.

Tumblr[1] has emerged as a leading social media platform with more than 260 million users, where user-generated content in the form of posts is shared

---

[1] tumblr.com; Acquired by Yahoo! in 2013.

© Springer International Publishing Switzerland 2016
V. Sugumaran et al. (Eds.): WEB 2015, LNBIP 258, pp. 222–227, 2016.
DOI: 10.1007/978-3-319-45408-5_26

with followers and can be further *reblogged*. Many companies use Tumblr as a channel to actively engage with customers and to deliver messages about new products or services.[2] In this paper, we empirically analyze the sharing behavior in Tumblr to test the significance of the two factors, messages and messengers, on the information diffusion process. Specifically, we analyze posts (i.e., messages) created by 102 company Tumblr accounts and track how they are reblogged.

Tumblr posts contain *textual* and *visual* information. For textual information such as text and tags, Latent Dirichlet Allocation (LDA) topic modeling [2] is used to extract meaningful features following its successful application in the management literature [9]. For visual information, a major challenge is how to incorporate such rich, but unstructured data into the analysis. Traditionally, processing raw visual data required significant engineering efforts and domain knowledge to extract meaningful features. This has started to change with the recent breakthrough of deep learning methods that have emerged as a powerful class of models giving dramatically improved performance on various prediction tasks [4]. Encouraged by these results, we employ deep learning features for images as a useful and robust representation. To the best of our knowledge, this paper is one of the first works to adopt deep learning approaches in the management literature. To capture the effect of messengers, we utilize social relationship of users and blogs. Specifically, we incorporate not only follow relationships, but also community structure based on the follow graph. We employ a community detection algorithm [3] to identify tightly connected clusters of users and examine how sharing behavior differs with the cluster structure.

2   Hypotheses Development

Attractive to consumers, multimedia content is frequently used in companies' social media posts [6]. Such contents tend to have a higher impact and deliver information more effectively in a short time to consumers. Thus, we expect that: **Hypothesis 1 (H1)** – *Company posts with photos (including gif images) or videos will receive more reblogs*. In addition, photos with different levels of complexity may have various advertising effects as shown in [7]. For example, photos with higher visual complexity can hinder the viewer's perception of an advertisement [8]. Thus, we hypothesize that: **Hypothesis 2 (H2)** – *Posts with less complex photos will receive more reblogs*.

It is shown that the initial injection points (the first rebloggers in our case) are important in the diffusion process [1]. Thus, we expect that: **Hypothesis 3 (H3)** – *Posts reblogged by users who have more followers will receive more reblogs*. For effective advertising, company posts should be targeted to the consumers with high interests in the content subject. We examine how content similarity between a company post and a user's previous contents influences subsequent reblogging behaviors: **Hypothesis 4 (H4)** – *Posts reblogged by users whose previous posts have high similarity with the focal posts will receive more reblogs*.

---

[2] https://www.tumblr.com/business.

3   Data and Variable Construction

As a social network service, Tumblr users can *follow* blogs that are of interest. This forms a directed *follower graph*, where nodes are users/blogs and edges represent follows. As a microblogging platform, Tumblr provides useful tools to create rich content. Using the follower graph and post content features, we analyze 3,773 posts created by 102 official company blogs for a one month period. In addition, we use post data of users who reblogged any of the company posts, which includes 1.8 million posts and 1 million images from 168 K blogs.

Topic Modeling for Text and Tags: To extract features from text data, we employ LDA topic model [2]. The basic assumption of LDA is that a given text document consists of a few latent topics and that the words in the document are the realization of the underlying topics. We collected text and tags from Tumblr posts as two separate corpora. We computed LDA with 50 topics for each corpus and use the topic distribution of each document (i.e., post) as textual features.

Deep Learning Features for Images: The majority of Tumblr posts are photos, thus images are important subjects of our analysis. Our dataset has more than 1 million images generated by companies and rebloggers. To extract useful and robust features from the images, we employ the deep learning approach proposed by [4], which constructs an 8-layer deep convolutional neural network. Each layer transforms the input image from the previous layer, where the transformation is learned from the data with the goal of accurately classifying objects in the image. For each image, the model outputs a 1,000 dimensional vector of confidence scores corresponding to different object categories.

Community Information: From the follower graph, we identify tightly connected communities using a widely used method called modularity optimization [3]. Modularity measures the difference between the expected and the actual number of intra-cluster edges. This partitioning algorithm proceeds in a top-down manner starting with the entire graph and stops when modularity does not increase by further partitioning the graph. With a follower graph of 76.86 million nodes and 2.27 billion edges, the algorithm identified 2,895 communities.

Variables: The dependent variable, num_reblog, is the total number of reblogs a company post receives through a direct reblogger. If user u reblogs company post p and three other users reblog p from u, then num_reblog = 4 for the $\langle u, p \rangle$ pair. There are 267,810 such pairs. Note that we use the standard deviation of feature vectors (*_sd) as a compact representation of how the topics (text and tags) or object categories (images) are distributed. That is, higher standard deviation values imply that the post focuses on a small number of topics/objects, whereas lower values indicate diverse topic/object coverage. We also compute the cosine similarity between the feature vectors, which is useful to analyze the homophily effects. Table 1 describes the variables used in the analysis.

Table 1. Variable description.

Variable name	Description
num_reblog	Number of reblogs received by a direct reblogger
{feat}_num	Number of {feat} in company post[a]
{feat}_sd	Standard deviation of {feat} for company post[a]
{feat}_sim	Cosine similarity of {feat} between company post and blog[a]
in/out_degree	Log-scale in/out degree of user blog
gif, videos	Company post contains gif/video
community	⟨company, user⟩-pair belongs to the same community
if_follower	User follows company blog

[a]feat = {image(photos),text(words),tags}

4 Empirical Analysis

We investigate how various post, blog, and user characteristics influence reblogging behavior. Due to potential unobserved characteristics, it is a challenge to find causal effects. To avoid potential endogenous issues, we use panel data fixed effects models in two different settings to separately examine each dimension.

Impact of Content Characteristics on Post Popularity: To capture unobserved blog heterogeneity, we add fixed effects for each company blog in the model. We use negative binomial regression as the dependent variable (num_reblog) has overdispersion. To measure the idiosyncrasy of a company post, we compute similarity (*_sim) between the focal post and the average post from the company's blog. Results are listed in Column 1 in Table 2. We observe that having more photos (photo_num) yield more reblogs supporting **H1**. In contrast, video posts get less reblogs (video). Another finding is that conspicuous images (image_sd) have positive impact on reblog counts, which supports **H2**.

Heterogeneous Influence of Content Characteristics: Next we investigate how the effect varies for companies with different characteristics. Companies use different marketing strategies based on their audience sizes. Thus we include interaction terms (*_int) between the number of followers for company blogs (in_degree) and post characteristics (video, gif, photo) in the model. Results are listed in Column 2 in Table 2. We observe that the coefficients of gif_int and photo_int are significantly negative, while the coefficients of gif and photo are significantly positive. This implies that, for less popular blogs, multimedia content in posts will increase attention from the audience. But this advantage diminishes for more popular blogs.

Publisher's Impact on Post Popularity: Publishers have different number of followers resulting in different levels of post exposure. To exclude potential

Table 2. Empirical results on reblogging behavior.

Variable	Negative binomial	Neg. binomial - interaction	Negative binomial	Logit
video	−0.13(0.10)	0.73(0.63)		
video_int		−0.09(0.07)		
gif	0.07(0.05)	1.56(0.38)**		
gif_int		−0.16(0.04)**		
photo_num	0.67(0.009)**	0.38(0.07)**		
photo_int		−0.03(0.008)**		
tags_num	−0.03(0.007)**	−0.04(0.007)**		
words_num	0.002(0.002)	0.001(0.001)		
image_sd	13.99(2.18)**	13.02(2.18)**		
text_sd	−0.38(0.39)	−0.29(0.39)		
tags_sd	0.63(0.37)	0.80(0.37)*		
image_sim	−0.037(0.075)	−0.025(0.075)	−0.055(0.063)	0.18(0.084)*
text_sim	0.001(0.08)	−0.027(0.08)	0.48(0.039)**	0.43(0.054)**
tags_sim	0.49(0.075)**	0.80(0.369)**	0.40(0.03)**	0.17(0.044)**
in_degree	0.19(0.017)**	2.18(0.26)**	0.69(0.003)**	0.72(0.005)**
out_degree	0.01(0.014)	0.01(0.014)	−0.086(0.03)**	−0.093(0.004)**
community			0.17(0.011)**	0.12(0.014)**
if_follower			0.12(0.02)**	0.061(0.023)**
Constant	−2.36(0.17)**	−2.98(0.21)**	−4.87(0.032)**	−3.90(0.106)**
Fixed effect	Company blog		Company post	
Observations	3,437		267,807	

endogeneity of post content, the next post-level analysis measures the impact of different rebloggers within the same posts. We apply post-level fixed effects to control unobserved post characteristics correlated with post popularity. Logit and negative binomial models are used as the distribution of reblog counts is highly skewed. Note that similarity is measured between the focal post and the average post from a user blog. Results are shown in Columns 3–4 in Table 2. We observe that a company post gets more reblogs when reblogged by users with similar content (*_sim) supporting **H4**. In addition, posts reblogged by users in the same community with the company blog (community) receive more reblogs. Lastly, we observe that rebloggers with more followers (in_degree), but less followers (out_degree) tend to yield more reblogs as in **H3**.

References

1. Banerjee, A., Chandrasekhar, A.G., Duflo, E., Jackson, M.O.: The diffusion of microfinance. Science **341**(6144) (2013)
2. Blei, D.M., Ng, A.Y., Jordan, M.I.: Latent dirichlet allocation. J. Mach. Learn. Res. **3**, 993–1022 (2003)
3. Blondel, V.D., Guillaume, J.-L., Lambiotte, R., Lefebvre, E.: Fast unfolding of communities in large networks. J. Stat. Mech. Theory Exp. **10**, P10008 (2008)
4. Jia, Y., Shelhamer, E., Donahue, J., Karayev, S., Long, J., Girshick, R., Guadarrama, S., Darrell, T.: Caffe: convolutional architecture for fast feature embedding. In: ACM International Conference on Multimedia, pp. 675–678 (2014)

5. Lee, D., Hosanagar, K., Nair, H.: The effect of social media marketing content on consumer engagement: evidence from facebook. In: SSRN, p. 2290802 (2014)
6. Liaukonyte, J., Teixeira, T., Wilbur, K.C.: Television advertising and online shopping. Mark. Sci. **34**(3), 311–330 (2015)
7. Pieters, R., Wedel, M., Batra, R.: The stopping power of advertising: measures and effects of visual complexity. J. Mark. **74**(5), 48–60 (2010)
8. Pieters, R., Wedel, M., Zhang, J.: Optimal feature advertising design under competitive clutter. Manage. Sci. **53**(11), 1815–1828 (2007)
9. Shi, Z., Lee, G.M., Whinston, A.B.: Towards a better measure of business proximity: topic modeling for industry intelligence. MIS Q. forthcoming
10. Shi, Z., Rui, H., Whinston, A.B.: Content sharing in a social broadcasting environment: evidence from twitter. MIS Q. **38**(1), 123–142 (2014)
11. Stieglitz, S., Dang-Xuan, L.: Emotions and information diffusion in social media-sentiment of microblogs and sharing behavior. J. Manage. Inf. Syst. **29**(4), 217–248 (2013)

The Impact of Patient Web Portals on Readmissions Through IT Based Reminder Systems

Yazan Alnsour[✉] and Jiban Khuntia

Business School, University of Colorado Denver, Denver, CO, USA
{Yazan.Alnsour,Jiban.Khuntia}@ucdenver.edu

Abstract. Patient web portals can play an important role in alleviating rising healthcare costs. Portals can help patients to get reminders about conducting diagnostic tests and care regime after a hospital admission to reduce subsequent readmissions. This research in progress has the focus to investigate the effect of portal based health reminders on the unplanned hospital readmissions. We argue reminders would have a reducing effect on readmissions with each view of diagnostic test reminders. An empirical strategy is proposed that would use an archival data set through a medical provider. We discuss the contributions and practice of our findings.

Keywords: Patient web portals · Reminders · Readmission · Digital health · Comorbidity

1 Introduction

The evidence of patient portals to improve health outcomes, care efficiency and utilization remains ambiguous in both practice and research equivocally (see reviews of studies on patient portals by [1, 2]. Although existing studies suggest that patient portals have positive effects on the usability aspects [3], but overall, there is a huge gap in the existing literature to explore the impact of patient portals, specifically in the context of its effectiveness for patients to manage their disease progression.

A distinct feature of a patient portal is the provision for doctors to send reminders to patients. A medical reminder shows the required test that the patient need to do. Reminders are shown in a screen that explain briefly each needed tests and highlight the urgency to take an action and conduct them. A patient will view the reminder as soon as he or she enters into the portal with a login and password based access. Once the test is conducted, the provider flags the test reminder as completed or been adhered to by the patient. Subsequently, the reminder is take of the patient's portal screen. Some examples of tests reminders that are shown to patients through a patient portal are albumin limit tests, blood sugar tests, uric acid tests, and ferritin levels tests for iron deficiency. Often multiple disease conditions lead providers to send multiple reminders via the portal that may have a bearing on patient's timely decision to conduct tests.

In this study, we explore how medical reminders reduce unplanned readmissions. We conceptualize a model (see Fig. 1) that posit for a reducing effect of reminders views (through number of clicks) on unplanned hospital readmissions. In addition, argue that

V. Sugumaran et al. (Eds.): WEB 2015, LNBIP 258, pp. 228–232, 2016.
DOI: 10.1007/978-3-319-45408-5_27

comorbidity and treatment adherence have complementing roles on this effect. We propose to empirically examine these relationships using an archival data set from a large primary care provider in the United States. We discuss the plausible practice implications and contributions of the findings.

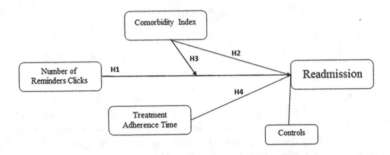

Fig. 1. Conceptual model

2   Theoretical Background and Hypothesis Development

We base our theoretical framework on prior research on information processing and individual decision making concepts. Researchers suggest that decision making process is a step-wise problem-solving activity terminated by a solution deemed to be satisfactory [4]. The involved step-wise process may be reason driven or emotion driven with either rational or irrational logic. Further, the process may be an information or cognition reliant approach, and possibly would include cost-benefit analyses to weigh available options [5].

Based on the theoretical frame work of information processing and decision making to conduct a test in the presence of a reminder, we propose a conceptual model for this study (see Fig. 1). The conceptual model posits that the number of clicks on the reminder will provide the health-contextual and informational cues to the patient. As the information (content) and contextual (health concerning) cues and subsequent phases for information processing activates a state of motivation or concern on the part of the patient [6, 7], the role of comorbid conditions will play a complementary role to either increase or reduce concerns and subsequent evaluations for conducting the test. We argue that if the patient receive reminders regarding comorbid conditions, then the patient would have a higher concern to respond to the reminders as soon as possible, and hence will manage his or her care in a better manner which will result in less hospital admissions.

2.1   Reminders Clicks and Readmission

With each click and view of the reminder, the patient undergoes a series of encoding and interpretation process, involving disease progression, accessibility and availability for the tests, and economic rationale to be able to afford the cost of medical diagnostic tests [7]. Some scholars advocate the benefits of an ambience or cue to enhance a patient's internal or intrinsic motivation to proceed towards the treatment for a disease [8]. Removing barriers to information, and the process of feeling empowered, may involve redefining and communicating the patient's role, enhancing his or her knowledge about the treatment options, health condition and status. Reminders being descriptive and highly informative (than a simple test name or ambiguity on tests), have a role to motivate the patient to conduct the tests. Indeed, as the patient views the reminder a number of times, his or her interpretation about the information relevant to the disease and treatment outcomes, information access, clarity of the process involved, and the psychological and economic rationale will lead him or her to conduct the associated test as soon as possible and that will lowers future possibilities of readmission [7]. Therefore, we hypothesize:

Hypothesis H1. Number of Reminders clicks is negatively associated with hospital readmission.

2.2 Comorbidity, Treatment Adherence Time and Readmission

Comorbidity indicates the presence of two or more diseases that are occurring with a primary disease. The perception of comorbidity are highly concerning to a patient, because comorbidity reflects that the patient's health is affected by multiple diseases and high rate of mortality. The state of disease affliction with two or more diseases would be a highly distressing situation for the patient [9], and he or she will have a high motivation to get a clarity about the situation. Research shows that with a high emotional situation in regards to health conditions, patients would seek a clear and reliable information [10]. Because the diagnostic tests will provide a clarity to the patient on his or her multiple disease situation, he or she would have a higher motivation to adhere to treatment and respond to medical reminders by conducting the required tests as soon as possible. Based on that, we hypothesize:

Hypothesis H2. Comorbidity index is positively associated with hospital readmission.

Hypothesis H3. Treatment Adherence Time is positively associated with readmission.
    While the presence of comorbid conditions provides an indication or cue to the presence of multiple diseases, the decision to conduct the tests depends on the psychological and rational cognitive and rational evaluation process. We argue that with each view of the medical reminder, the patient goes through a cycle of such evaluation process. The evaluations subsequently trigger a rationalization process, in which the patient considers the attributes associate with the decision making process, such as time, cost or affordability options, accessibility and other similar factors [11]. These factors together would also reflect on the patient's tangible and intangible resource constraints to conduct the tests, out of which, purely economic factors include reducing insurance premium costs or patient co-pay would play an important role towards the decision to do a test. Thus, we argue that in case of comorbidity, the multiple disease condition may be alarming for the patient to increase his or her anxiety to take an expedited action.

Each click and view of the medical reminder has the potential to reinforce the deferred test decision or the expedited test decision. Based on these arguments, we hypothesize:

Hypothesis H4. Comorbidity index complement number of clicks to reduce readmission.

3 Proposed Method

We consider Glycated hemoglobin test (known as A1C) conducted to diagnose a pre-diabetic or diabetic condition as the main reminder for this study. A1C test is recommended for both (a) checking the blood sugar control in people who might be pre-diabetic and (b) monitoring blood sugar control in patients with more elevated levels, termed diabetes mellitus [12]. High levels of A1C may also indicate the patient has a risk of hypoglycemia or advanced stage of diabetes leading to other health disease and issues, such as renal failure. Normal levels of glucose in blood produce a normal amount of Glycated hemoglobin, and as the average amount of plasma glucose increases, the fraction of Glycated hemoglobin increases in a predictable way. This serves as a marker for average blood glucose levels over the previous months prior to the measurement. For comorbidity we use Charlson Comorbidity Index. The index predicts the mortality for a patient who may have comorbid conditions. Each condition is assigned with a score. Higher scores indicating greater comorbidity. The number of Readmissions is a continuous variable. Therefore, we use ordinary least squares (OLS) estimation in our models. We will estimate two equations corresponding to the direct and interaction sets of hypotheses to be tested.

$$READM = \beta_{10} + \beta_{11}(CLICKS) + \beta_{12}(COMBD) + \beta_{13}(TAT) + \beta_{1c\eta}Control_\eta + \varepsilon_1 \quad (1)$$

$$READM = \beta_{20} + \beta_{21}(CLICKS) + \beta_{22}(COMBDI) + \beta_{23}(CLICKS \times COMBDI) + \\ \beta_{24}(TAT) + \beta_{2c\eta}Control_\eta + \varepsilon_2 \quad (2)$$

At this current research in progress stage, we have collected data from a provider. We have found preliminary support for the hypotheses. We plan to conduct a set of robustness tests to validate our findings.

4 Discussion

The goal of this study is to explore how medical reminders via patient portals reduce unplanned readmissions. We will empirically examine the effect of reminders view (through number of clicks) on unplanned hospital readmissions using an archival data set from a large primary care provider in the United States for patients, who were recently admitted to a hospital and received and viewed reminders through a patient portal. To our knowledge this is will be the first study in the existing information systems literature that explores the impact of medical reminders via patient web portals on reducing unplanned readmissions within 30 days.

This study contributes to the emerging stream of research in information system surrounding the impact and transformation brought in with health information technologies (HITs) [10, 13].

The study has some limitations. First, currently the study is planned to limit to only for one provider. Although this helps to enhance internal validity of the study and control for many cross-provider effects, it may limit the generalizability of the study. Similarly, this study is conducted considering one focal medical reminder, future studies may replicate for other reminders.

The study provides implications for effectiveness of patient web portals through IT based reminder. Further, we contribute to the existing information systems and healthcare research streams of research in exploring the impact of a patient portals on enhancing care delivery, and highlight the importance of patient portal and alert systems as emerging patient-centered health information technologies.

References

1. Ammenwerth, E., Schnell-Inderst, P., Hoerbst, A.: The impact of electronic patient portals on patient care: a systematic review of controlled trials. J. Med. Internet Res. **14**(6), e162 (2012)
2. Goldzweig, C.L., et al.: Electronic patient portals: evidence on health outcomes, satisfaction, efficiency, and attitudes a systematic review. Ann. Intern. Med. **159**(10), 677–687 (2013)
3. Britto, M.T., et al.: Usability testing finds problems for novice users of pediatric portals. J. Am. Med. Inform. Assoc. **16**(5), 660–669 (2009)
4. Kahneman, D., Tversky, A.: Choices, values, and frames. Am. Psychol. **39**(4), 341 (1984)
5. Schacter, D.L., Gilbert, D.T., Wegner, D.M.: Psychology. Worth Publishers, New York (2010)
6. Hilligoss, B., Rieh, S.Y.: Developing a unifying framework of credibility assessment: construct, heuristics, and interaction in context. Inf. Process. Manag. **44**(4), 1467–1484 (2008)
7. Sillence, E., et al.: How do patients evaluate and make use of online health information? Soc. Sci. Med. **64**(9), 1853–1862 (2007)
8. McAllister, M., et al.: Patient empowerment: the need to consider it as a measurable patient-reported outcome for chronic conditions. BMC Health Serv. Res. **12**(1), 157 (2012)
9. Bowman, G., Watson, R., Trotman-Beasty, A.: Primary emotions in patients after myocardial infarction. J. Adv. Nurs. **53**(6), 636–645 (2006)
10. Anderson, C.L., Agarwal, R.: The digitization of healthcare: boundary risks, emotion, and consumer willingness to disclose personal health information. Inf. Syst. Res. **22**(3), 469–490 (2011)
11. Marteau, T.M., Ashcroft, R.E., Oliver, A.: Using financial incentives to achieve healthy behaviour. BMJ **338**, b1415 (2009)
12. American Diabetes Assoiciation. http://www.diabetes.org/
13. Kohli, R., Kettinger, W.J.: Informating the clan: controlling physicians' costs and outcomes. MIS Q. **28**, 363–394 (2004)

Do Hedonic and Utilitarian Apps Differ
in Consumer Appeal?

Bidyut Hazarika[✉], Jiban Khuntia, Madhavan Parthasarathy,
and Jahangir Karimi

Business School, University of Colorado Denver, 1475 Lawrence Street,
Denver, CO 80202, USA
{bidyut.hazarika,jiban.khuntia,madhavan.parthasarathy,
jahangir.karimi}@ucdenver.edu

Abstract. This research in progress suggests that hedonic and utilitarian mobile applications (apps) differ in consumer appeal. We operationalize consumer appeal through addiction, frustration and consumer value perception. The interplay of frustration in the presence of a set of addicted customers is interesting, as many apps may have an addicted consumer baser-specifically for games, communications of or entertainment apps. Consumer frustration may act as a negative complementing factor to consumer addiction, and this substitution effect is argued to be higher for hedonic products than utilitarian products. Hypotheses are drawn and a methodology is suggested. Possible implications and contributions are discussed.

Keywords: Consumer addiction · Consumer frustration · Digital products · Apps · Hedonic · Utilitarian · Consumer evaluation

1 Introduction

Continuous development in mobile technologies and related applications (apps) has emerged to converge in the apps market space. Starting with a handful of apps in 2008, the Apple Store has more than a million apps for sale in 2015. Similar trends are observed in the Android and Microsoft apps stores as well. Many companies are using mobile apps as traditional trade channels. Revenue from app users is projected to increase from $10.3 billion in 2013 to $25.2 billion in 2017 [5].

In this study, we propose to explore how hedonic and utilitarian apps differ in consumer appeal. We posit that consumer appeal is reflected by addiction, frustration and value perception, amongst other factors. We ask the research question how the relationships of consumer addiction and frustration on consumer evaluation is differentiated by hedonic and utilitarian apps. Based on prior studies [13], we denote the term hedonic to those apps that "aim to provide self-fulfilling value to the user"; whereas utilitarian apps "aim to provide instrumental value to the user". A conceptual model is suggested with possible avenues for data collection and research execution. Possible contributions and implications are discussed.

© Springer International Publishing Switzerland 2016
V. Sugumaran et al. (Eds.): WEB 2015, LNBIP 258, pp. 233–237, 2016.
DOI: 10.1007/978-3-319-45408-5_28

2   Conceptual Development and Hypotheses

Existing information systems has differentiated the hedonic and utilitarian use of information technology, through the lens of usage behavior as a result from a reasoned appraisal of pre-adoption beliefs of the IT artifact [9]. A few studies have extended the hedonic and/or utilitarian lens to mobile computing (e.g. [11]) and other general IT products (e.g. [8]).

Addiction is the use of something for relief, comfort, or simulation, and which often continues in part, due to cravings when it is absent and has affected people's life in various ways and some people may require treatment [14]. When someone becomes dependent on a certain type of technology and have to use it repeatedly, it becomes a behavioral addiction [4]. The existence of such addiction to digital artifacts has been proven by Turel and Serenko [10].

Addiction is a trait when a consumer gets obsessed with a certain product and keeps using it repeatedly over and over again. In case of apps, if a consumer keeps checking an app repeatedly to check its content, we can classify them as addiction. This repeated use of the app leads to eventual addiction of the consumer. Consumer addiction indicates not only the liking of consumers for an app due to its appeal, function or on the attribute to meet a specific unique need; but also that the app has the potential to be sustained in the market due to a set of addicted consumers. For example, a fitness app would have a good appeal for the consumer segment that want to stay fit, and hence, would have generated a set of consumers who use and get *addicted to it*. These addicted consumers would then derive higher benefits, be testers, will work towards the improvement of the app by providing feedback and in general help in mitigating the aspects related to the frustration technology issues. The features such as an app having a great design, integration, functionality contributes towards the addiction. An app in a work place works smoothly as advertised can lead to greater productivity and utilization from its workers resulting in good evaluations, leading to positive word of mouth and better customer evaluation.

Consumer frustration is the result of the failure a consumer faces when a technology fails to achieve its desired task. In an individual level, consumer frustration can be defined as the negative emotion due to an unsuccessful use of a technology [1]. In case of an app, the main cause of consumer frustration is inaccurate description, not enough engagement, not user friendly, crash and other usage problems. Consumer frustration leads to dissatisfaction, and loss of self-efficacy and eventually lead to consumer uninstalling the app and leaving a bad review. Consumer frustration may also lead to disruption in workplace, slow functionalities and not using the app [6] or also may lead to high levels of anxiety and anger on part of the user [12] eventually leading the user to stay away from the technology [7]. As pointed out by Guchait and Namasivayan [2], all the above mentioned factors create a bad response for the technology in the market.

Consumer frustration deals with the negative emotions that individuals feel with the use of technology. In the context of apps, if an app doesn't perform as it is supposed to, it leads to consumers discontinuing the app, or being dissatisfied by the app. The performance of the app in this study is based on the factors such as crash, design, usage, integration and functionality issues. Prior studies note that consumer frustration leads to loss of the users' self-efficacy and subsequent personal dissatisfaction. For example, when the user used an app for a game or to stream a movie, and the app fails or crashes; then it leads to discontinuation of the game or the movie, leading to dissatisfaction of the user. Similarly, an apartment finder app, or a business networking app, if do not work properly as desired, lead to failure of the business and loss of productivity and revenue. Often the resulting negative word-of-mouth from dissatisfied users will lead to less adoption of the app which in turn will lead to decrease the consumer evaluation of the app, finally leading to the death of the app. Prior studies (e.g. [2, 3]) note that the frustration felt by consumers when shared in a public forum creates a negative image of the product in the market.

Based on the discussions in the existing literature, as elaborated in previous paragraphs, we develop a conceptual model for this study. Figure 1 below shows the proposed conceptual model. We posit that addiction and frustration have direct effects on consumer evaluation of an app. However, these direct effects being not so surprising, we do not hypothesize those, and focus on the differentiating hypotheses of hedonic and utilitarian apps.

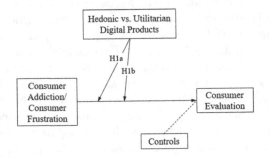

Fig. 1. Conceptual model

When people decide to use an app, it can be due to two values: utilitarian and hedonic behavior. Utilitarian behavior is more relational and task related. We classify utilitarian apps as those apps that are used by consumers in an effective manner. For example, apps that are downloaded to track finances, medical records are all classified as utilitarian apps: apps that are used to achieve certain tasks. Hedonic apps on the other hand are apps that are used for fun and playfulness vs. task completion. Thus, we posit that hedonic apps are mostly used by consumers for entertainment value such as: arousal, heightened involvement, perceived freedom, fantasy fulfillment and escapism.

Hedonic apps will have more success as they are more likely to be used more often compared to utilitarian apps. The degree of addiction will also be high for these apps as well. If hedonic apps works seamlessly, the consumer evaluation will be much higher as people would have greater evaluation for these apps. In case of utilitarian apps, people use them only to achieve certain task and not for its enjoyment value. So, if a utilitarian app crashes or have technical glitches, consumer will be more forgiving compared to a hedonic app. Based on these arguments, we hypothesize that:

H1a: *The positive effect of consumer addiction on consumer evaluation is higher for hedonic apps than utilitarian apps.*

H1b: *The negative effect of consumer frustration on consumer evaluation is higher for hedonic apps than utilitarian apps.*

3 Proposed Method and Discussion

This research in progress suggests to collect secondary data for several weeks, code the variables using text mining and analyze the data using econometric methods. We expect to find support for the hypotheses. We also plan to conduct a set of robustness tests to validate our findings.

This study aims to report the findings on whether consumer frustration has a negative association on consumer evaluation, whether the interaction of addiction and frustration has differentiating effects on consumer evaluation for hedonic and utilitarian apps.

The findings would inform managers to focus on design aspects, such as app designs can be oriented to have more "hedonist" design aspects, even though they have utilitarian value. Further, the study would inform managers that technical design and integration plays a highly valuable role in consumer acceptance of the app. Finally, as much as an app's release is important, taking feedback from consumers and tracking evaluation regularly (almost on a daily basis) is critical for an app's success.

In terms of research contributions, this study as of now is the first one to explore the comparative effects of consumer behavioral factors associated with apps adoption; with a set of nuanced variables such as addiction, frustration associated with hedonic and utilitarian apps. Further, indirectly this study identifies the important role of design and evaluation factors associated with IT business value, specifically in the apps market context. Moreover, it also informs to the vast literature in the marketing on the product feature or aspects that influences consumer evaluation of products.

In conclusion, the objective of this study is to explore to what extent consumer addiction and consumer frustration influences consumer evaluation of apps; and how these effects differ across utilitarian and hedonic apps. We propose to analyze secondary data for android apps and to validate the model and hypotheses. This study contributes to the research on the digital business strategy for apps markets.

References

1. Bessière, K., Newhagen, J.E., Robinson, J.P., Shneiderman, B.: A model for computer frustration: the role of instrumental and dispositional factors on incident, session, and post-session frustration and mood. Comput. Hum. Behav. **22**(6), 941–961 (2006)
2. Guchait, P., Namasivayam, K.: Customer creation of service products: role of frustration in customer evaluations. J. Serv. Market. **26**(3), 216–224 (2012)
3. Hazarika, B., Karimi, J., Khuntia, J., Parthasarathy, M.: Consumer Frustration and Consumer Valuation Shift for Mobile Apps: An Exploratory Study (2015)
4. Holden, C.: 'Behavioral' addictions: do they exist? Science **294**(5544), 980–982 (2001)
5. IDC.: Worldwide and U.S. Mobile applications download and revenue 2013–2017 forecast: The app as the emerging face of the internet (2013)

6. Lazar, J., Jones, A., Hackley, M., Shneiderman, B.: Severity and impact of computer user frustration: a comparison of student and workplace users. Interact. Comput. **18**(2), 187–207 (2006)
7. Lenhart, A.: The Ever-Shifting Internet Population: A New Look At Access and the Digital Divide. Pew Internet & American Life Project, Washington, D.C. (2003)
8. Lin, C.P., Bhattacherjee, A.: Extending technology usage models to interactive hedonic technologies: a theoretical model and empirical test. Inf. Syst. J. **20**(2), 163–181 (2010)
9. Mallat, N.: Exploring consumer adoption of mobile payments–a qualitative study. J. Strateg. Inf. Syst. **16**(4), 413–432 (2007)
10. Turel, O., Serenko, A., Bontis, N.: User acceptance of hedonic digital artifacts: a theory of consumption values perspective. Inf. Manag. **47**(1), 53–59 (2010)
11. Wakefield, R.L., Whitten, D.: Mobile computing: a user study on hedonic/utilitarian mobile device usage. Eur. J. Inf. Syst. **15**(3), 292–300 (2006)
12. Wilfong, J.D.: Computer anxiety and anger: the impact of computer use, computer experience, and self-efficacy beliefs. Comput. Hum. Behav. **22**(6), 1001–1011 (2006)
13. Wakefield, R.L., Whitten, D.: Mobile computing: a user study on hedonic/utilitarian mobile device usage. Eur. J. Inf. Syst. **15**(3), 292–300 (2006)
14. Young, K.S.: Internet addiction: the emergence of a new clinical disorder. Cyber Psychol. Behav. **1**(3), 237–244 (1998)

Erratum to: E-Life: Web-Enabled Convergence of Commerce, Work, and Social Life

Vijayan Sugumaran[1]([⊠]), Victoria Yoon[2], and Michael J. Shaw[3]

[1] Department of Decision and Information Sciences, Oakland University,
Rochester, MI, USA
sugumara@oakland.edu
[2] Virginia Commonwealth University, Richmond, VA, USA
[3] Beckman Institute for Advanced Science and Technology, University of
Illinois at Urbana–Champaign, Urbana, IL, USA

Erratum to:
V. Sugumaran et al. (Eds.):
E-Life: Web-Enabled Convergence of Commerce, Work,
and Social Life, LNBIP 258,
https://doi.org/10.1007/978-3-319-45408-5

In the original publication of this book the conference numbering in the subtitle was incorrect. It was the 14th Workshop but the subtitle erroneously stated "15th Workshop". This has now been corrected.

The updated original online version of this book can be found at
https://doi.org/10.1007/978-3-319-45408-5

Author Index

Printed in the United States
By Bookmasters